B1.4 Variation

 What is variation?

In 2009, a BBC camera crew discovered a new species of giant rat, deep in the jungles of Papua New Guinea. It was 82 cm long and not afraid of humans. While scientists are agreeing its scientific name, it is being called the Bosavi woolly rat (after the place in which it was found).

A A Bosavi woolly rat

1 a What is variation?
b What variation is there between the rats in Figure B?

Differences in characteristics are called **variation**. There is variation within a species but much more variation between different species. This variation within a species can make classification difficult.

B These rats are all of the same species but show great variation. Great variation within a species can be confusing and make classification difficult.

Scientists must make sure that any 'new' organism is not just a hybrid or due to variation in a known species. They do this by finding more than one of the organisms.

From the Bosavi woolly rats that have been found, scientists can tell that they are closely related to other 'hairy rats' because they share many characteristics. However, Bosavi woolly rats have sufficient variations in those characteristics to show that they are a new species.

Keys

To identify different species, you can use a **key**. Figure C shows a key for rats.

2 a What evidence is there that the woolly rat is a different species to the rats in Figure B?
b How might further evidence be collected to show that they are different species?

3 Why do scientists need to find more than one example of a 'new' organism?

4 Use the key to identify which rat species is shown in Figure B

Statement:	Next step:
1 The animal has a thick, bushy tail.	squirrel
The animal has a tail that is covered with layers of skin, which look like scales.	**go to 2**
2 It has a band of hair that is longer than the rest of the fur.	crested rat
Its fur is all the same length.	**go to 3**
3 It has a short tail.	groove-toothed rat
It has a long tail.	**go to 4**
4 Its tail is shorter than its body.	brown rat (*Rattus norvegicus*)
Its tail is longer than its body.	black rat (*Rattus rattus*)

C To use the key, pick the correct statement from the first box and follow the 'next step' instructions. Carry on doing this until you reach a name.

The importance of classi

Accurate classification using the b
• easily identify existing species a
• see how organisms are related
• identify areas of greater and less

Biodiversity is a measure of the to
To count the species you need to
if species are very similar to one a
system, the easier identification be

Biodiversity is important because
(e.g. foods, medicines). The more
have, both now and in the future.
recovering from natural disasters

Many biologists think that areas of
are the ones that need the most t
them because this will result in a g

D Biodiversity hotspots (shown in red

5 How does classification make it
6 Why is it important to protect b
7 Design a key to identify some farm
species and from the same species.

Learning Outcomes

1.8 Explain why binomial classification is
conservation efforts
1.9 Explain how accurate classification ma
1.10 Construct and use keys to show how
HSW 1 Explain how scientific data is collecte

34

Progress questions can be used to check understanding as you work through the course.

Learning outcomes are taken straight from the specification to make it clear what you need to know for the exam.

Maths skills boxes appear throughout the Student Book to give you opportunities to refresh your maths skills.

These pages give you an investigation task, allowing you to complete a specification practical. These tasks can be used as practice Controlled Assessments.

The specification practical is listed in the Learning outcomes box.

edexcel
advancing learning, changing lives

EDEXCEL GCSE SCIENCE

Science Student Book

Series Editor: Mark Levesley

Richard Grime

Miles Hudson

Penny Johnson

Sue Kearsey

Damian Riddle

Nigel Saunders

ResultsPlus authors:

Pauline Anderson

Mark Grinsell

Mary Jones

Ian Roberts

David Swann

A PEARSON COMPANY

Contents

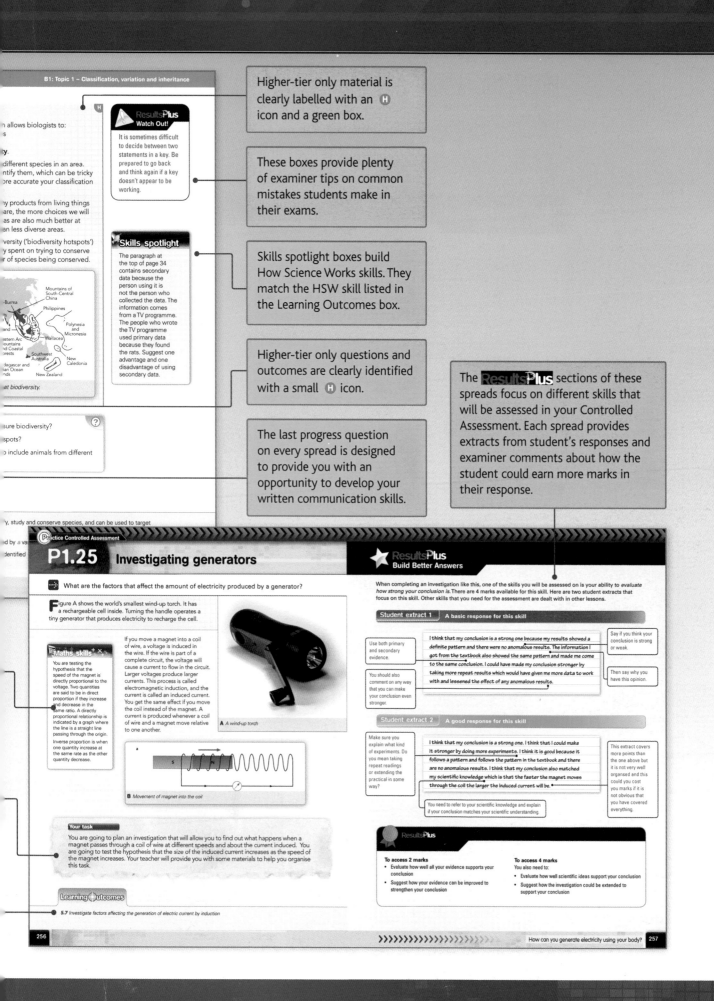

Higher-tier only material is clearly labelled with an H icon and a green box.

These boxes provide plenty of examiner tips on common mistakes students make in their exams.

Skills spotlight boxes build How Science Works skills. They match the HSW skill listed in the Learning Outcomes box.

Higher-tier only questions and outcomes are clearly identified with a small H icon.

The ResultsPlus sections of these spreads focus on different skills that will be assessed in your Controlled Assessment. Each spread provides extracts from student's responses and examiner comments about how the student could earn more marks in their response.

The last progress question on every spread is designed to provide you with an opportunity to develop your written communication skills.

B1: Topic 1 – Classification, variation and inheritance

ResultsPlus Watch Out!

It is sometimes difficult to decide between two statements in a key. Be prepared to go back and think again if a key doesn't appear to be working.

Skills spotlight

The paragraph at the top of page 34 contains secondary data because the person using it is not the person who collected the data. The information comes from a TV programme. The people who wrote the TV programme used primary data because they found the rats. Suggest one advantage and one disadvantage of using secondary data.

Practice Controlled Assessment

P1.25 Investigating generators

What are the factors that affect the amount of electricity produced by a generator?

Figure A shows the world's smallest wind-up torch. It has a rechargeable cell inside. Turning the handle operates a tiny generator that produces electricity to recharge the cell.

Maths skills

You are testing the hypothesis that the speed of the magnet is directly proportional to the voltage. Two quantities are said to be in direct proportion if they increase and decrease in the same ratio. A directly proportional relationship is indicated by a graph where the line is a straight line passing through the origin.

Inverse proportion is when one quantity increase at the same rate as the other quantity decrease.

If you move a magnet into a coil of wire, a voltage is induced in the wire. If the wire is part of a complete circuit, the voltage will cause a current to flow in the circuit. Larger voltages produce larger currents. This process is called electromagnetic induction, and the current is called an induced current. You get the same effect if you move the coil instead of the magnet. A current is produced whenever a coil of wire and a magnet move relative to one another.

A A wind-up torch

B Movement of magnet into the coil

Your task

You are going to plan an investigation that will allow you to find out what happens when a magnet passes through a coil of wire at different speeds and about the current induced. You are going to test the hypothesis that the size of the induced current increases as the speed of the magnet increases. Your teacher will provide you with some materials to help you organise this task.

Learning Outcomes

5.7 Investigate factors affecting the generation of electric current by induction

---

ResultsPlus Build Better Answers

When completing an investigation like this, one of the skills you will be assessed on is your ability to evaluate how strong your conclusion is. There are 4 marks available for this skill. Here are two student extracts that focus on this skill. Other skills that you need for the assessment are dealt with in other lessons.

Student extract 1   A basic response for this skill

Use both primary and secondary evidence.

You should also comment on any way that you can make your conclusion even stronger.

I think that my conclusion is a strong one because my results showed a definite pattern and there were no anomalous results. The information I got from the textbook also showed the same pattern and made me come to the same conclusion. I could have made my conclusion stronger by taking more repeat results which would have given me more data to work with and lessened the effect of any anomalous results.

Say if you think your conclusion is strong or weak.

Then say why you have this opinion.

Student extract 2   A good response for this skill

Make sure you explain what kind of experiments. Do you mean taking repeat readings or extending the practical in some way?

I think that my conclusion is a strong one. I think that I could make it stronger by doing more experiments. I think it is good because it follows a pattern and follows the pattern in the textbook and there are no anomalous results. I think that my conclusion also matched my scientific knowledge which is that the faster the magnet moves through the coil the larger the induced current will be.

This extract covers more points than the one above but it is not very well organised and this could you you marks if it is not obvious that you have covered everything.

You need to refer to your scientific knowledge and explain if your conclusion matches your scientific understanding.

ResultsPlus

To access 2 marks
- Evaluate how well all your evidence supports your conclusion
- Suggest how your evidence can be improved to strengthen your conclusion

To access 4 marks
You also need to:
- Evaluate how well scientific ideas support your conclusion
- Suggest how the investigation could be extended to support your conclusion

How can you generate electricity using your body?

There is a practice exam paper for both the Foundation and Higher tiers in each unit.

Hydrocarbon fuels

5. Many hydrocarbons are used as fuels.

   (a) Below are the structures of molecules of two hydrocarbons.

   hexane   octane

   Name the two elements that are combined together in molecules of alkanes. (2)

   (b) Hexane is a liquid. During complete combustion, oxygen reacts with hexane to form carbon dioxide and water.

   (i) Write the word equation for the reaction. (1)

   (ii) What different product would form during the combustion of hexane if there was a restricted amount of oxygen available? (1)

   (iii) State two factors that make a good fuel. (2)

   (c) The amount of energy released when a fuel burns can be determined using the apparatus shown in the diagram.

   thermometer

   clamped calorimeter containing 100 g of water

   spirit burner

   You have been given two hydrocarbon fuels to investigate, hexane and octane. Explain, in detail, how you could use the apparatus safely to see which fuel gives out the most heat energy. (6)

190

Rocks

1. The photographs show samples of a sedimentary rock, a metamorphic rock and an

   Rock P   Rock Q

   (a) Rock P is sedimentary. What evidence could you **see** when examining a it is sedimentary?

   (b) Rock Q is metamorphic. Metamorphic rocks are formed when:

      A   magma cools
      B   layers of small rock fragments build up
      C   existing rocks are changed by heat and pressure
      D   sedimentary rocks are melted and solidified

   (c) Rock R is igneous. It contains crystals. Explain how magma forms igneou sized crystals.

   (d) Limestone is a sedimentary rock composed of calcium carbonate, $CaCO$ forms calcium oxide, CaO, and a gas when heated strongly.

      (i) Complete the

         $CaCO_3 \rightarrow$ ——

      (ii) The United Ki disadvantage

All exam practice questions are written by examiners.

Each exam section has 2 extended writing questions where you can practise answering this type of question. The extended writing questions are worth 6 marks and are always the last part of questions 4 and 5.

The question parts are colour coded to indicate what grades they can access:
- Orange means you can access grades G-D (Foundation tier) or D-B (Higher tier)
- Light green means you can access grades E-C or B-A*
- Dark green covers the whole grade range, G-C or D-A*.

The Build Better Answers pages present an extended writing question along with three different student answers to the question – a level 1, a level 2 and a level 3 answer.

ResultsPlus
Build Better Answers

Here are three student answers to the comments around and after them.

Question   Light an

Claire carried out an investigation to find out how light affected the growi plants.

She placed 8 cress seeds into each o dishes A, B, C and D. Claire shone lig onto dishes A, B and C from differen directions. Dish D was kept in the da Claire left the dishes for one week. T diagrams show the results of Claire's investigation.

Claire controlled several variables wi she carried out her investigation. De which variables Claire could control, she could control these variables and they need to be controlled. (6)

Student answer 1   Extract t

Good – two important variables have been mentioned.

She should make
temperature. She
they all get the sa
amount of water a

Examiner summary
Everything in this answe
been controlled. The ans
could control the amount
variables need to be con

106

Exam practice spreads provide a large bank of the new question types you will encounter in your exams.

The Be the Examiner spread offers a variety of past exam questions along with examiner comments, advice on how to improve an answer and common mistakes that students have made in the past.

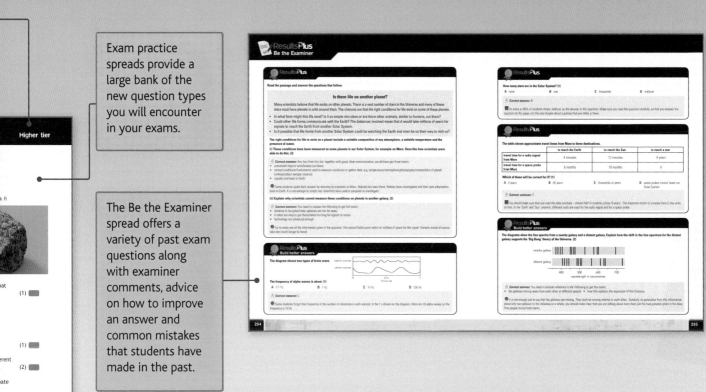

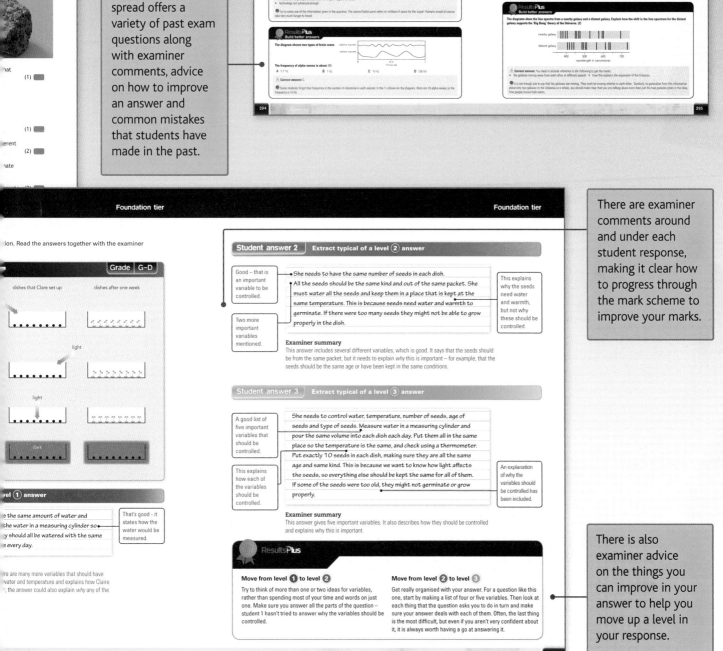

There are examiner comments around and under each student response, making it clear how to progress through the mark scheme to improve your marks.

ion. Read the answers together with the examiner

Grade | **G–D**

dishes that Clare set up   dishes after one week

light

light

dark

Student answer 2 — Extract typical of a level ② answer

Good – that is an important variable to be controlled.

She needs to have the same number of seeds in each dish.
All the seeds should be the same kind and out of the same packet. She must water all the seeds and keep them in a place that is kept at the same temperature. This is because seeds need water and warmth to germinate. If there were too many seeds they might not be able to grow properly in the dish.

Two more important variables mentioned.

This explains why the seeds need water and warmth, but not why these should be controlled.

Examiner summary
This answer includes several different variables, which is good. It says that the seeds should be from the same packet, but it needs to explain why this is important — for example, that the seeds should be the same age or have been kept in the same conditions.

Student answer 3 — Extract typical of a level ③ answer

A good list of five important variables that should be controlled.

She needs to control water, temperature, number of seeds, age of seeds and type of seeds. Measure water in a measuring cylinder and pour the same volume into each dish each day. Put them all in the same place so the temperature is the same, and check using a thermometer. Put exactly 10 seeds in each dish, making sure they are all the same age and same kind. This is because we want to know how light affects the seeds, so everything else should be kept the same for all of them. If some of the seeds were too old, they might not germinate or grow properly.

This explains how each of the variables should be controlled.

An explanation of why the variables should be controlled has been included.

Examiner summary
This answer gives five important variables. It also describes how they should be controlled and explains why this is important.

ResultsPlus

Move from level ① to level ②
Try to think of more than one or two ideas for variables, rather than spending most of your time and words on just one. Make sure you answer all the parts of the question – student 1 hasn't tried to answer why the variables should be controlled.

Move from level ② to level ③
Get really organised with your answer. For a question like this one, start by making a list of four or five variables. Then look at each thing that the question asks you to do in turn and make sure your answer deals with each of them. Often, the last thing is the most difficult, but even if you aren't very confident about it, it is always worth having a go at answering it.

el ① answer

e the same amount of water and the water in a measuring cylinder so y should all be watered with the same e every day.

That's good - it states how the water would be measured.

re are many more variables that should have ater and temperature and explains how Claire , the answer could also explain why any of the

There is also examiner advice on the things you can improve in your answer to help you move up a level in your response.

Why study science?

If you asked your science teacher why it's important to study and understand science, they may say, 'Everything around you is science.' You might expect a scientist to say that, but there's truth in it.

Just think of what your daily life would be like without advances in science. Even over the last 100 years, the achievements of science have benefited the human race in so many areas: in transport (cars, planes), in medicine (anaesthetics, organ transplants, cures for many diseases, drugs), in electronics (computers, mobile phones), in materials (fabrics such as Gore-Tex ® and Kevlar ®) and in agriculture (fertilisers, pesticides, genetic modification) to name but a few.

However, sceptics will say these advances have come at a price. Alfred Nobel, whose will funds the Nobel Prizes, invented dynamite for the mining industry, yet it rapidly became part of the armaments industry. And a hundred years ago, no one had heard of global warming or the ethical arguments over gene therapy. But in order to understand whether these sceptics are right, it's important that you learn some of the science behind these issues.

So why study science? Firstly, there is the excitement of discovery. Although you may not discover anything new yourself, you'll be able to experience some of the 'buzz' of the great scientists by working out some key ideas for yourself.

Science also gives you the chance to do some practical work: to get your hands dirty with experiments. You can investigate living organisms, create chemical reactions that give off energy or make colour changes, or build a physical model of a phenomenon.

However, learning about science in the new millennium is about more than that – and this is where we return to our sceptics. You need to be able to make informed decisions about how science benefits your life and what to do in situations where science has posed as many questions as it has answered. Much of what you might see in the media presents one side of an argument – and is often designed as a piece of journalism, high on shock value and low on facts! Knowing some facts yourself will enable you to look critically at how science is presented in the media.

It's also important to realise the importance of science to your future. Scientists work in so many fields to try and improve the life and health of the planet. Some will be in research, perhaps working on the cure for cancer or HIV / AIDS, or helping make the chips to power the next generation computer-game technology. Others may be working in industries trying to develop new energy sources or new fabrics and materials. But you'll also find science specialists working as weather forecasters, television researchers, lawyers, medical specialists, teachers, writers, architects, journalists and in many other areas. Increasingly, employers look for applicants with a science background because they know they will have logical, enquiring minds.

Finally, there's one other excellent reason to study science: it's fun! We hope that this book will help you throughout GCSE Science to see the interest, relevance and enjoyment of the subject.

Good luck!

Damian Riddle
Assessment Team Leader,
Science
Edexcel

The units

GCSE Science is made up of a series of small units and you'll take a number of these units over the two or three years you study the course.

To get your GCSE in Science, you'll need to take four units. Of these, three will be written exams: one each in Biology (B1), Chemistry (C1) and Physics (P1). You'll find out more about these exam papers on the next page. The fourth unit is a piece of coursework, called a Controlled Assessment (CA). If you want to find out more about this, look at pages 18–25 of this book.

If you go on to GCSE Additional Science, the pattern is the same: three written exam papers in Biology (B2), Chemistry (C2) and Physics (P2); and a piece of Controlled Assessment.

Each science also has an Extension Unit if you're taking separate sciences (sometimes called 'triple science'). So if you're doing GCSE Biology, you'll also do three written exam papers (B1, B2 and B3) and a Controlled Assessment.

Confused? It's probably easiest to understand how this works by looking at the diagram.

As you can see, each GCSE works exactly the same: three exams and a Controlled Assessment.

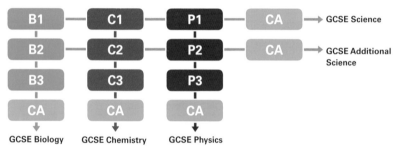

What will you study in your course? Let's think just about the three units that make up the GCSE Science.

Unit B1

This Biology unit is split into three topics.

In Topic 1, you will:
- explore the general characteristics of animals and plants, in particular vertebrates and organisms that can survive in extreme environments
- learn how to classify organisms, study basic variation and principles of inheritance and appreciate Darwin's theory of evolution by natural selection.

In Topic 2, you will:
- learn how humans detect and respond to changes in their environments, including the role of hormones and the nervous system.

In Topic 3, you will:
- look at how the body is affected by drugs and disease-causing organisms
- learn how scientists have developed antibiotics and antiseptics
- study nutrient cycles and think about how chemical substances produced by human activities pollute our planet.

Unit C1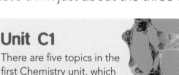

There are five topics in the first Chemistry unit, which cover the following areas:

- how the Earth and its atmosphere have evolved and how human activity can have an impact on them

- the types of rock in the Earth, especially calcium carbonate and how it reacts to form other compounds which are useful in everyday life, like being used to reduce harmful emissions

- how acids react to give useful products and how electricity is used to make new substances

- how metals are extracted from their ores and used to make alloys

- how crude oil is a source of many substances in our world, from petrol to plastics to Persil!

Unit P1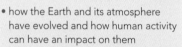

The Physics unit is split into six smaller topics. In these topics you will:

- learn about light and lenses, how telescopes led to the development of our knowledge of the Solar System and the evolution of stars and the Universe

- learn how light fits into the electromagnetic spectrum and how other components of the spectrum are used

- discover how the Universe may have evolved from the Big Bang, how we measure it and how we study the planets and stars of our Solar System

- explore waves, from communication between whales, to scanning of unborn babies, to earthquakes

- learn how an electric current is generated and how electricity can be transmitted over large distances

- investigate energy transfers in common appliances and in our atmosphere.

In the next few pages, we'll take an in-depth look at the exams you'll take in GCSE Science.

This will help you get familiar with what the papers look like and will also give you some information on the sorts of questions that you'll be asked, plus the way in which the examiners use key words to test your knowledge and understanding.

Assessment Objectives

Many students think that examiners have the job of trying to catch them out or that examiners set out to write exam papers which are too hard. Nothing could be further from the truth! Examiners try hard to write questions that allow you to show everything you have learned during your GCSE course. However, they also have to make sure that any exam paper is fair for you, past students and future students. The papers need to be the same difficulty and test the same skills.

To help with this, all papers are targeted to test specific skills called Assessment Objectives. They sound a bit complicated, but here's what they mean:

Assessment Objective	The jargon	What it means
AO1	Recall, select and communicate your knowledge and understanding of science.	Essentially, AO1 questions will ask you to write down facts that you have learned.
AO2	Apply skills, knowledge and understanding of science in practical and other contexts.	AO2 questions may ask you to apply what you've learned to new situations, or to practical contexts – it's about showing skills you've learned, rather than repeating facts.
AO3	Analyse and evaluate evidence, make reasoned judgements and draw conclusions based on evidence.	AO3 questions will be about how well you can use data, say what graphs show, or think about arguments with two sides, such as ethical issues.

Your exam papers will be written so that the balance of these AOs is roughly 45% AO1, 35% AO2 and 20% AO3. The Controlled Assessment task will be about 50% AO2 and 50% AO3.

Types of question

Your exam paper will always consist of six questions. The questions will examine different areas of the course which you have studied. Each question will start with some straight-forward question parts, which will slightly increase in difficulty until the end of the question. The six questions themselves will also be 'ramped' slightly in difficulty: this means that Question 6 will be a little more challenging than Question 1. However, each question will start with a question part that puts you at ease and settles you into the question as a whole.

The first two questions will be worth around 8 marks, the next pair around 10 marks and the final pair around 12 marks. Each question could contain the following types of question:

- **Multiple choice**. These ask a question and usually provide four possible answers: A, B, C and D. Some multiple choice questions may ask you to choose words from a box to complete a sentence or to draw lines between statements.

- **Open response**. Like multiple choice, these will usually be worth 1 mark, but you will need to write down your own answer, rather than choose from a selection of answers provided.

- **Short answers**. These questions, worth 2 or 3 marks, will ask you to write slightly longer answers (perhaps three or four lines). Occasionally, at Higher tier, some questions – most likely those involving calculations – may be worth 4 marks.

Two questions in the paper, usually questions 5 and 6, will also have a piece of extended writing.

- **Extended writing**. These questions are worth 6 marks. They will require you to write at greater depth and the questions will be slightly more open to give you the chance to express yourself. In these questions, you will also be assessed on the quality of your written communication.

There is more detail on each type of question, and how to answer them, on pages 14 and 15.

The language of the exams

An exam paper contains precise words. It's important that you understand exactly what the words are asking you to do so that you can answer the question quickly and accurately.

The words are called 'command words' and it's useful for you to know what they mean. Some are simple words designed to allow you to give simple answers, whereas others are more complex terminology, asking you to write at greater length and give more detail.

Let's have a look at the command words you might encounter on a GCSE Science paper. These are in order of complexity below, i.e. they get more difficult.

Give, Name, State	These questions are asking for short answers – often only a few words – but with precise use of scientific terminology.
Complete, Select, Choose	In these questions you are usually asked to fill the gaps in sentences, using words that are given to you.
Draw, Plot	These questions will ask you to put data onto a graph, or draw a diagram. Make sure that your work is accurate and that you use labels, if appropriate.
Calculate, Work out	In GCSE Science, you'll need to use your mathematical skills in solving equations and performing simple calculations. This type of question will test you on these skills.
Describe, Use the graph	In these questions, try to use scientific terminology concisely and accurately. You may be asked to describe a particular process – in which case, remember to think about putting your answers in a logical sequence. Alternatively, you may be asked to describe the trend in a graph – here you will get more marks for using data that you have read from the graph.
Suggest	These questions are often asking you to apply what you have learnt to a new situation, or to put together different facts you have learnt in order to answer the question.
Explain	These questions will be worth 2 or 3 marks. You need to give reasons for your answer and not to simply repeat information given in the question. Imagine that any question that starts 'Explain...' is really asking you 'Why...?' – that should help you understand the depth needed in your answer.
Evaluate, Discuss, Compare	These command words will often be found in the 6-mark extended writing questions. Look carefully at exactly what the question is asking you to do. In many cases, it will be asking you to look at 'pros' and 'cons' surrounding a particular argument, or to give a balanced answer, looking at different aspects of a topic. 'Compare' questions will ask you to look at areas of similarity and difference between two ideas, or two sets of data.

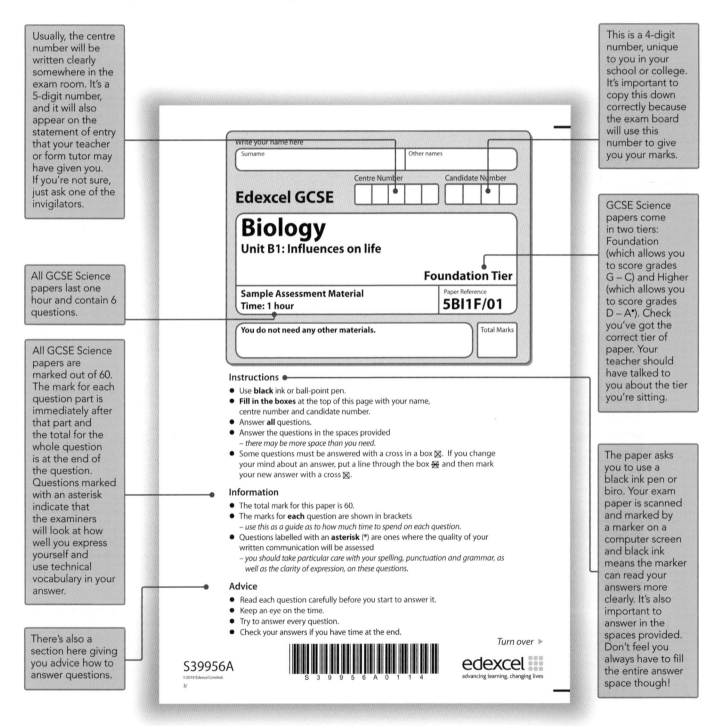
Knowing what to expect: the front page

We know it can feel a bit scary sitting in the exam room with your GCSE paper in front of you, waiting to start. If you know what the paper looks like beforehand and what some of the text on it means, hopefully it won't seem so daunting.

Usually, the centre number will be written clearly somewhere in the exam room. It's a 5-digit number, and it will also appear on the statement of entry that your teacher or form tutor may have given you. If you're not sure, just ask one of the invigilators.

This is a 4-digit number, unique to you in your school or college. It's important to copy this down correctly because the exam board will use this number to give you your marks.

All GCSE Science papers last one hour and contain 6 questions.

GCSE Science papers come in two tiers: Foundation (which allows you to score grades G – C) and Higher (which allows you to score grades D – A*). Check you've got the correct tier of paper. Your teacher should have talked to you about the tier you're sitting.

All GCSE Science papers are marked out of 60. The mark for each question part is immediately after that part and the total for the whole question is at the end of the question. Questions marked with an asterisk indicate that the examiners will look at how well you express yourself and use technical vocabulary in your answer.

The paper asks you to use a black ink pen or biro. Your exam paper is scanned and marked by a marker on a computer screen and black ink means the marker can read your answers more clearly. It's also important to answer in the spaces provided. Don't feel you always have to fill the entire answer space though!

There's also a section here giving you advice how to answer questions.

Write your name here

Surname

Other names

Centre Number

Candidate Number

Edexcel GCSE

Biology
Unit B1: Influences on life

Foundation Tier

Sample Assessment Material
Time: 1 hour

Paper Reference
5BI1F/01

You do not need any other materials.

Total Marks

Instructions

- Use **black** ink or ball-point pen.
- **Fill in the boxes** at the top of this page with your name, centre number and candidate number.
- Answer **all** questions.
- Answer the questions in the spaces provided
  – *there may be more space than you need.*
- Some questions must be answered with a cross in a box ☒. If you change your mind about an answer, put a line through the box ☒ and then mark your new answer with a cross ☒.

Information

- The total mark for this paper is 60.
- The marks for **each** question are shown in brackets
  – *use this as a guide as to how much time to spend on each question.*
- Questions labelled with an **asterisk** (*) are ones where the quality of your written communication will be assessed
  – *you should take particular care with your spelling, punctuation and grammar, as well as the clarity of expression, on these questions.*

Advice

- Read each question carefully before you start to answer it.
- Keep an eye on the time.
- Try to answer every question.
- Check your answers if you have time at the end.

Turn over ▶

S39956A
©2010 Edexcel Limited.
3/

S 3 9 9 5 6 A 0 1 1 4

edexcel
advancing learning, changing lives

Knowing what to expect: a sample question

This is called the stem of the question. It lets you know what topic the question is on. In some cases, the stem gives some important information to set the scene for the question.

Each question will contain what examiners call a stimulus. Most often, this will be a photograph or a diagram that should help set the context for the question. In some cases, the stimulus might be some data from an experiment or a short piece of text.

Here's an example of one of the command words that we looked at on page 11. It tells you what to do in your answer.

Examiners use bold text to highlight something they want you to notice. In many cases, it will be a word which you will be asked to define or it will tell you the number of answers you are expected to give.

The number of marks for the question part is given on the right-hand side of the exam paper, just under the question part. It's always a good idea to look at how many marks the question part is worth. This will give you some idea of how much you are expected to write, what sort of depth the examiners are looking for and the length of time you should spend on the question part.

Multiple choice questions are laid out differently – we'll look at these in more detail on the next page.

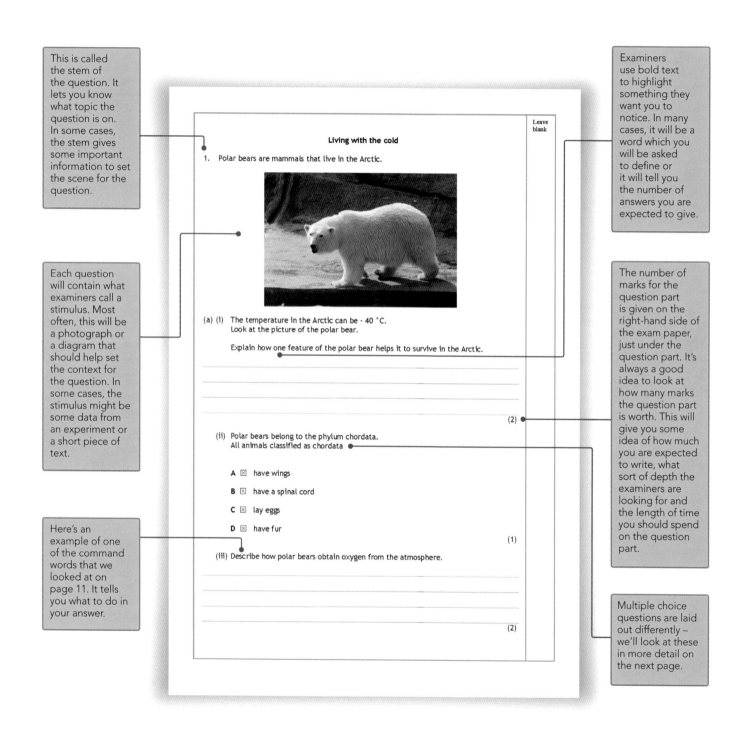

Leave blank

Living with the cold

1. Polar bears are mammals that live in the Arctic.

(a) (i) The temperature in the Arctic can be - 40 °C.
Look at the picture of the polar bear.

Explain how **one** feature of the polar bear helps it to survive in the Arctic.

_____

_____

_____

(2)

(ii) Polar bears belong to the **phylum** chordata.
All animals classified as chordata

A ☒ have wings

B ☒ have a spinal cord

C ☒ lay eggs

D ☒ have fur

(1)

(iii) Describe how polar bears obtain oxygen from the atmosphere.

_____

_____

_____

(2)

On pages 10 and 11, we looked at the types of questions that would come up in your GCSE Science exams. Let's look at each one in a bit more detail, so you understand how to score as many marks as you can.

Multiple choice questions

The most common type of multiple choice question will present you with four alternative answers, one of which is correct. For example:

> Which of these is an addictive substance contained in tobacco?
>   A a cannabinoid
>   B carbon monoxide
>   C nicotine
>   D tar

The best way to answer multiple choice questions is to cover up the four possible responses when you read the question. Think of your answer and then look at the four possible answers: hopefully yours will be among them!

If your answer isn't there – or if you couldn't come up with an answer when you looked at the question to start with – then you have to start looking at the answers one by one and eliminating ones that you know are wrong. Hopefully, you'll be able to eliminate three answers and be left with the correct one. It may be that you're only able to eliminate one or two wrong answers – but at least you can make a more educated guess. The key thing with a multiple choice question is this: don't leave it blank. You can't lose marks if you get it wrong and you may actually guess right!

There are similar questions where you're given the answers as part of the question. These might be selecting words from a box to complete sentences, or drawing lines to join words in one column to those in another. The same principle applies to these types as to the standard 'ABCD' multiple choice questions – look at the question before looking at the answers to see if you already know the answer.

Open response and short-answer questions

These questions might be 1 mark questions, often asking you to recall information. Or they may be worth 2 or 3 marks, if you're being asked for two reasons, or two factors.

Some examples of 1 mark questions might include:

> a Give one form of naturally-occurring calcium carbonate.
> b Which waves in the electromagnetic spectrum have the longest wavelength?

Other types of 1 mark questions might ask you to label or complete diagrams. Some may be of the **Suggest** type, asking you to use information that you already know and combine it with information in the question to come up with an answer.

Remember that these questions are only worth 1 mark – so don't spend too much time going into a great amount of detail. Keep your answer clear and concise. You don't have to answer in full sentences, so you can save time here.

Longer-answer questions

Each question on the paper will have some questions worth 2 or 3 marks. Generally, there will be more of these longer-answer questions towards the end of the exam paper. These types of question will include the following:

Writing or completing equations. Make sure with these that you're clear whether the answer needs a word equation or a chemical equation. Don't forget to look for clues in the question – you'll often find that the words or formulae you need are in the question.

Drawing graphs. Different questions will ask you to do different things with a graph. Read the question carefully to check if you're meant to label axes or draw your own axes. Be careful when you plot the points – there are marks for the accuracy of your plotting. Finally, check to see if you're required to draw a line of best fit – and remember, lines of best fit don't have to be straight lines, they can be curves!

Performing calculations. Two key tips for calculations. Firstly, always show your working. If you make a slip with your calculator (and end up with the wrong final answer), you can often get marks for your working if the examiner can see that you just made a small slip. Secondly, check to see if you need to give units in your answer.

Describe or **Explain** questions. For example, a question like:

> Describe the path taken by a nerve impulse from the receptors to the effectors.

These questions are designed to be a bit more difficult than some of the other types we've seen. However, there's no reason why you shouldn't score well on them. It's always best to spend a few moments planning your answer before starting to write. Make sure you understand exactly what you're being asked to do – you want to keep your answer relevant. Do think about the number of marks that the question is worth: if it's a 3 mark question, then you need to make three different points.

Extended writing questions

There is one final type of question that you'll face in your examination paper – the extended writing question. That may sound a bit daunting, but these questions aren't designed to be really difficult or to set traps for you. Instead, they give you more time and space to show the knowledge and skills you have picked up during your GCSE course. These questions will be written to test you on a variety of different skills.

The other important thing to realise about these questions is that they're not just aimed at A* students. The questions are designed to be open enough to allow all students to be able to write something.

On the next two pages we'll look at the sort of topics that might form the basis of an extended writing question.

Practical-based questions

Much of the time you spend in science lessons, you're doing practical work. Some extended writing questions will therefore ask you about the practical work that you've done. The questions may ask you about how you would set up an investigation or about how a particular practical investigation gives us information about scientific theories.

Opinion-based questions

Many people think that science is all about hard facts and certainties but many areas of science still lead to uncertainty. In some cases, this is because scientists genuinely don't know all the answers or because data that they collect can be interpreted in different ways. In other cases, the discoveries that scientists have made have led to moral or ethical dilemmas.

Opinion-based questions will therefore ask you to use what you have learnt to back up your own opinion or to show that you understand two sides of a particular argument.

Knowledge-based questions

There will be some parts of your GCSE Science course where you will learn quite a lot of detail. Sometimes, 1 or 2 mark questions on these topics mean you can only write about some of the things you know about the topic, rather than all of it. Equally, there are times when the examiners want to ask a general question to allow you to show a variety of things you know or to bring in knowledge across the whole subject.

Hence, some extended writing questions will ask you for more factual responses. This might seem a bit daunting, but the questions may sometimes be phrased in such a way as to allow a variety of answers. There will often be some prompt questions to help you get started.

How they're marked

The extended writing questions are always worth 6 marks. So if you've read the advice given for other questions types, you're probably thinking about the amount that you may have to write, and you're thinking that you have to make sure you make 6 credit-worthy points.

Well, you're half right – the whole point about these questions being called 'extended writing' is that there is the opportunity for you to write at greater length. Note that in many questions, you'll be able to provide diagrams or tables as part of your answer, so don't think it's all about writing essays! But you don't have to think about making six points each time. These questions will be marked with two ideas in mind – one is known as 'levels' and the other as 'QWC'.

Let's look at **levels** first. The mark scheme for each question contains a long list of possible points – the key thing to remember is that you're not expected to write all of these! The examiners will see lots of examples of answers, and they'll use this experience to place your answer in one of three levels: very good (Level 3), good (Level 2) and less good (Level 1). Their judgement will be based on factors such as the number of points you've made, the balance in any argument you've made or how well you've appreciated your practical work. Each level corresponds to 2 marks. Level 1 is 1–2 marks, Level 2 is 3–4 marks, and Level 3 is 5–6 marks. So once your answer is placed in one of the levels, how is the final mark decided?

The other deciding factor is **QWC**. This stands for Quality of Written Communication. What the examiners will look for is your ability to use proper scientific words correctly and use proper spelling, punctuation and grammar, your ability to put your answer in a logical order and how well you express yourself.

How to answer

Let's have a look at a possible question:

> Discuss the advantages and disadvantages of using biofuels instead of petrol for cars.

The question relates directly to a particular part of the specification – the statement about biofuels being renewable but taking up land that could be used for growing crops. However, your answer wouldn't have to involve that. You could, for example, talk about what factors make a good fuel and use this to compare biofuel and petrol; or you could write about the amount of carbon dioxide given off by each fuel and hence talk about the greenhouse effect. Or if you'd learned a lot about the effects of acid rain, you could even write about sulfur impurities in petrol causing this, whereas biofuels (which don't contain sulfur) don't give the same problems.

The key, therefore, is to treat each question as a way for you to display some of the knowledge you've learned, rather than just a black-and-white question with one set answer.

In terms of your exam technique, the most important things to do are: a) think about all the different areas of what you've learned that you can use in your answer, and b) to spend some time planning your answer. The two extended questions are worth a total of 12 marks between them – a fifth of the marks on the paper – so you should spend a reasonable amount of time on them, probably about 6–8 minutes on each. And remember that this is one area where you need to be careful with the logical presentation of your answer. Finally, read the question again before answering – did you notice that it asked for advantages and disadvantages? Remember, therefore, to think about both pros and cons in your answer.

Let's look at a possible answer to this question:

> Biofuel is a renewable fuel. It is often made from fermenting the sugar in sugar beet to produce ethanol. Petrol, however, is not renewable – it is a fossil fuel and, at the rate humans are using it, it will soon be used up. Petrol, however, does give more energy out when it burns than biofuels do. So, although biofuels may be an answer to the problems of energy needs, we'll need to burn a lot of them to get our energy. That means that a lot of land needs to be used to grow the sugar crop – and that means less land for growing food for people or livestock. Many people see biofuels as being "green". It is certainly true that, unlike petrol, they don't contain sulfur impurities, so there is no sulfur dioxide produced when they burn, so no acid rain. Biofuels do, of course, produce carbon dioxide when they're burnt, just like petrol, and this is a powerful greenhouse gas. But, some of the effect of this may be offset as, to grow the next crop of sugar beet, carbon dioxide will be taken in by the plants photosynthesising.

Here is how an examiner responded to this student answer:

This answer looks at some pros and cons about using petrol. Importantly, it also looks at some negatives of using biofuels – such as land use – as well as seeing the positives. This means it should be a top-level response. The presentation of the answer is good, with accurate spelling and grammar, and the answer is logically structured. It should achieve a mark in the top level.

There is an aspect of your course where we try to assess you on your ability to handle practical work and use the results of experiments. This section is known as Controlled Assessment.

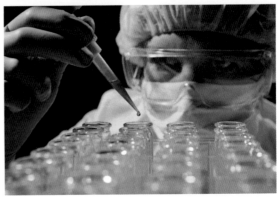

If you've got older brothers and sisters or friends in years above you at school, it's quite likely that they'll have told you horror stories about coursework and how much time it took up when they were studying for their GCSEs.

Well, Controlled Assessment is, essentially, a form of coursework, but one which is very different from the coursework used in the past. Hopefully, you'll see that many of the differences are positive ones.

The major differences are:

- In Controlled Assessment, the task is set by the exam board and changes every year. This means that it is the same task for all students, which is much fairer.

- In Controlled Assessment, parts of the work have to be completed in class time and you won't be able to take work home. This means that it is easier to guarantee that it is your own work and means that students who have family members who are good scientists aren't able to get their family to help with their work.

- Controlled Assessment takes up less time – especially because you do not have to do write-ups for several homeworks.

How important is Controlled Assessment?

Controlled Assessment is worth 25% of your GCSE – in other words, it has the same proportion of marks as a written exam.

How many Controlled Assessment tasks do you need to do?

You have to submit one Controlled Assessment mark for GCSE Science and another one if you go on to GCSE Additional Science. This can come from any of the three sciences: Biology or Chemistry or Physics. It is possible that your teachers may complete more than one Controlled Assessment task with you – if this is the case, you submit your best mark. In some cases, you may be able to combine marks for different Controlled Assessments in order to get a higher mark. Your teacher will be able to give you some advice on this.

If you are taking Biology, Chemistry and Physics as separate GCSEs, then you will need one Controlled Assessment for each science.

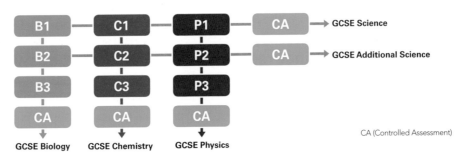

What sort of practicals will the exam board choose?

The specification for GCSE Science includes a series of suggested practicals. The Controlled Assessment could be set on one of these pieces of practical work, or it may be set on a related experiment. This is another example of us trying to ensure that the Controlled Assessment is as fair as possible. Having asked you to study some pieces of practical work in your course, it seems fair that one of these pieces of work – or something closely related to those pieces of work – will be used as the Controlled Assessment, as all candidates will have the same previous experience of the practical being used.

How will you prepare?

Preparation for the Controlled Assessment is obviously going to be very important. Hopefully, your teacher will give you lots of opportunities to practise the skills you need in order to do well in this section of the course. That's another reason why we've included the suggested practicals in each unit – so that you can use these experiments to help you get better at planning experiments, collecting data and making conclusions based on the data.

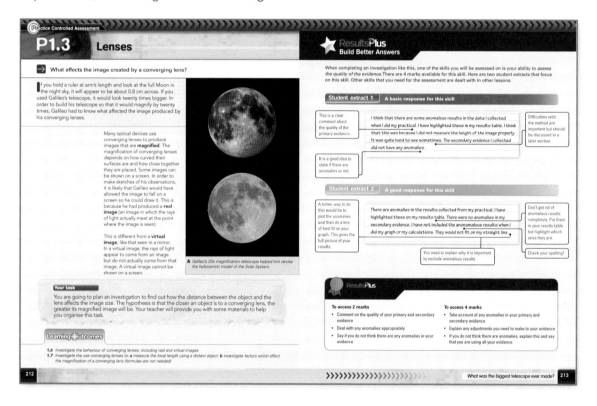

If you've had a flick through this textbook before reading this section, then you'll notice that each of the suggested practicals has a double–page spread in the book and that each of these concentrates on a different area of practical skills. You may do some of these suggested practicals as 'practice' Controlled Assessments to help you prepare for your real Controlled Assessment. Obviously in your real Controlled Assessment you will have a different investigation.

What do you have to do in a Controlled Assessment?

As far as possible, the idea of the Controlled Assessment task is to try and assess how well you have picked up key practical skills during your science course. This means that we have split up each Controlled Assessment into three distinct sections: Part A (planning), Part B (observing) and Part C (concluding).

Planning. In the planning section, you will be given a prediction (or, as scientists often call it, a hypothesis) to test. You'll also be given some information about the sort of experiment you should be doing to test the prediction. Remember, the Controlled Assessment will often be based on one of the suggested practicals that are in the specification – so you should have a good idea of what the experiment that you're trying to plan is like!

For GCSE Additional Science and for the separate sciences, you'll be asked to come up with your own hypothesis or idea to test.

Observing. In this section, you will carry out your plan and collect some data. You'll also research some secondary evidence. Your teacher may give you a pre-prepared plan to follow, rather than using your own. Don't worry if you think your practical skills are a bit basic, or if you know that you work very slowly and carefully when you do a practical. No–one is going to be standing over you to watch exactly what you're doing! We can assess how well you do an experiment by looking at the data that you get.

Concluding. Here, there are some key skills we're looking for: how well you present your data and draw relevant graphs; how you process information in order to make a conclusion; and, finally, whether you think the evidence that you have gathered during your practical supports the prediction or hypothesis you were aiming to test. Your conclusions will need to be based on your data and the secondary evidence that you collected.

It should be pretty clear that the Observing and Concluding sections go together – you can't analyse data if you've not collected any! But the Planning section is separate, so you may end up doing a Planning task based on one practical and the Observing and Concluding tasks on a different practical.

When will you do a Controlled Assessment?

That's up to your teacher – he or she will give you some notice that you'll be taking one of these tasks, so you'll have time to prepare. Doing well in a Controlled Assessment relies on you having picked up a certain number of skills in planning and analysing – so you'll probably be at least halfway through your GCSE course before you do your first piece of assessed work.

How will the real Controlled Assessment work?

It's possible that you'll do a practice before you do the real thing – or that you may do more than one real piece (because your teacher only has to submit your best mark – so if you do more than one, your best mark counts).

As you know, the Controlled Assessment is split into three parts and it's possible that you'll do the different parts at different times. Indeed, it's possible that you'll do the first part (the planning section) from a different Controlled Assessment task than the other two parts.

Why are practicals important?

One of the key things that marks science out as different to other school subjects is that it is practical.

It's important to realise that scientists do practical work not as a way to liven things up, but because science cannot progress without scientists making observations, collecting data and then using that information to back up (or disprove) a theory or hypothesis.

The philosopher Democrates is credited as first coming up with the idea that all substances are made of small particles called atoms, and that atoms cannot be broken up (in Greek, *atomos* means *indivisible*). Although modern atomic theory bases itself on this two-and-a-half-thousand-year-old idea, modern scientists have shown – practically – that this is not quite true. Indeed, harnessing the energy released by splitting the atom was one of the defining moments of the 20th century and, although humans haven't always used this technology for good, this is a source of energy on which we can rely as fossil fuels begin to be used up.

Throughout history, important discoveries in science have been linked to practical work and observations. By doing practical work and collecting data, scientists are able to put forward theories to explain the data produced. Sometimes, of course, the theory comes first; but the practical work is always there to help give evidence to show the theory is correct.

To think of examples from each science, Harvey could never have proposed the theory of blood circulation in humans without having performed experiments and dissections; Humfrey Davy would never have discovered and isolated so many chemical elements without using electricity to break up substances by electrolysis; and where would Newton's theory of gravity have been without the simple – if unplanned – experiment of dropping an apple on his head?

When designing this GCSE course, one aim has been to encourage your teachers to give you better access to practical work than has often been the case in the past. We think that practical work is very important for your enjoyment of science and, because it's also a great way to learn, we've tried to design our GCSE courses to put practical work back into them.

Part A – Planning

Your teacher will issue you with a student brief. This student brief will give you a bit of background, particularly telling you something about the experiment. It might be that the experiment you're doing is similar to one you've already encountered in your course. For GCSE Science, you'll also be given a prediction or a hypothesis to test.

The planning task is under limited control – this means that you can discuss ideas with people in the class and do some other work as preparation. It's possible that you'll spend some time in class with your teacher discussing ideas which may be important to help you plan. Otherwise, you may be asked to do some research for homework to help you prepare for the planning section. You may be given some prompt questions – some questions for you to think about as you do your preparation.

Some key things for you to consider will be:

- **The apparatus you need to do the experiment.** Think about what you're trying to measure and which apparatus will do this best – and be prepared to justify your choice of equipment. Remember that it's often better to provide a labelled diagram of apparatus rather than a list!

- **How many readings to take.** Here, you want to consider the range of readings that you wish to take as well as whether you want to repeat any readings. Again, be prepared to justify the decisions you come to.

- **How to control factors in the experiment.** As well as thinking about how you're going to change the factor you're investigating, you also need to think about how you're going to make sure that other factors stay the same – in other words, how you make sure that the experiment is a fair test.

- **Any relevant safety factors.** Try to make sure that these are relevant to the experiment you're actually doing and not just general things like "tie your hair back".

- **The scientific theory** which can be used to justify your prediction.

The writing-up of your plan should take place in lesson time – depending on the length of your science lessons, you may be given more than one lesson, but the work will be collected in from you at the end of each lesson.

There are slightly different ways in which the planning exercise will be given to you. You'll already have had the briefing document and your teacher may feel, especially if you've been given some prompt questions, that you've got enough information to write your plan. Some teachers may issue you with the prompt questions again, in written form, just to remind you. Alternatively, you may be given something that looks a bit more like a question paper, with the prompt questions laid out one by one, with blank space left for you to answer each one. Whichever form you get, remember the key questions above and try to write a logical, concise plan that covers all of those points.

Hopefully, you'll get a chance to carry out your plan, so make sure that it's clear to follow and has all the necessary detail.

What makes a good plan?

Let's say that the hypothesis you're given is:

"The greater the mass of fuel burnt, the greater the rise of temperature in some water heated by the burning fuel."

The following is one possible plan to test the hypothesis. It is not a bad plan, but can you see where it can be improved?

"I'm going to measure the rise in temperature when I burn 3 different fuels. I will put the burning fuel underneath a beaker of water and use the burning fuel to heat the water. I'll use a thermometer to measure the temperature of the water before the experiment and again once I have burnt the fuel. I'll measure out $100cm^3$ of water each time using the scale on the side of the beaker. I'll use a balance to measure the mass of the fuel burnt. The water being heated in the experiment might get quite hot, so I will have to take care when I handle the beaker containing hot water."

The list of apparatus is not very clear – what is being used to hold the fuel, for example? Also, no justification is given about the apparatus named – especially with regard to how well they make the measurements you need. So, is a beaker really a good piece of apparatus for measuring volume – rather than a measuring cylinder? And how accurate are the thermometer and the weighing balance?

Some factors are controlled – the volume of water is mentioned – but others are not mentioned e.g. using the same beaker each time. If someone else was doing this experiment, they'd need to know when to stop the experiment – this could be when a set mass of fuel has been burned or when there is a set temperature rise in the water. This is an important variable.

There's also no indication of some practical details to improve how well the experiment works e.g. stirring the water as it is heated, or stopping the fuel from evaporating.

It's clear how many fuels are being used – three – but is this a wide enough range of fuels to provide evidence for the hypothesis? There's also no mention of doing the experiment more than once in order to check that the results are correct and can be averaged.

Finally, there is a good, relevant safety feature mentioned; although other risks – especially of the burning fuel – have not been mentioned.

Part B – Observations

In most cases, your Part B task will follow the Planning task, and you will perform the experiment that you have just planned. However, your teacher may give you a new plan to carry out at this point if you have problems with the plan you wrote.

In some cases you may just complete the planning task because your teacher may choose to take the Planning task from one area of the course e.g. from Biology; and your Observations task from another e.g. Chemistry.

Remember, though, that the Part C (Conclusions) task will be based on the Observations task – so it is important that you collect a good set of data.

In many ways, the Observations task is the most straightforward – you will be following your plan or worksheet for the practical work you have to carry out. There are some things for you to consider when you carry out the practical and collect data or observations.

How many readings will you take for each experimental trial you do?

- If you're taking more than one reading, how are you going to check to see that your results each time are concordant? (Concordant is the scientific way of saying that the results agree with each other).

- You may also need to decide on a range of readings to take. For instance, if the plan tells you to set up a circuit and to vary the voltage in the circuit and measure the current, then you may have to decide on a range of voltages to use e.g. 2V, 4V, 6V, 8V, 10V and 12V.

How are you going to record your observations?

- Think here of whether you need any units in the data you are recording, and how many significant figures or decimal places you need to record.

The Observations task is also under limited control. This means that you may work in a group to collect data; and also discuss the range and number of readings you will take within the group. If you are working in a group, it's important that you take turns to do the stages in the experiment – don't rely on one person to do all the practical work and one person to do all the recording!

The final part of the Observations task will be for you to collect secondary evidence. Secondary evidence is often results from a similar experiment or an experiment testing a similar hypothesis. You'll be given some guidance on where to look for your secondary evidence.

Once you have collected data and observations, your teacher will probably ask you to hand these in. This is because you will need this information in the next part of the Controlled Assessment task and so it's very important that it is not lost, or taken home and accidentally left there!

Part C – Conclusions

In order to do the Conclusions task of the Controlled Assessment, you'll have to have completed the Observations task. Don't worry if you've been away when the Observations task was done – your teacher should be able to provide you with some results taken by other students in your class.

This part of the task is under high control. This means that you will be expected to work entirely by yourself. So if you were working in a group in the Observations task, make sure that the data you collected is written in a form that you can understand. It's a good idea to make sure that, if you worked in a group of two for Observations, that you make two copies of the data you collected.

Just as in the Planning task in Part A, there will be a series of prompt questions to help you with this part of the Controlled Assessment. Again, there will be different formats in which these questions may be given to you – if you are in doubt, ask your teacher.

Don't forget the secondary evidence you researched in Part B. You'll need to use this information, along with the data you have collected in Part B, in order to complete Part C.

The key areas for you to concentrate on as you look at the data or observations you have collected are:

- **How you are going to process and present your results**. Most of the time, you'll want to put your results into a table, so think about column headings and units as you draw your table. With many experiments, you'll have data that you can plot to make a graph – but is it best to have a bar chart or a line graph?

- **How well the data you have collected supports the prediction or hypothesis**. You need to be able to use your knowledge of some scientific principles in order to justify the conclusion you make here.

- **An evaluation of the experiment that you did**. Here, you need to think about two things: how the set-up of the experiment worked well and allowed you to collect data of a high quality; and secondly, if there were some areas where the set-up was not helpful. Try to think about how easy the experiment was to do, but also think about how well the equipment you used worked and how good the results you collected were. When you come to think about things which were not good about the experiment, come up with some suggestions about things that you might like to change in order to make it easier to collect results and to make the results more accurate. Lastly, consider what effects any weaknesses in the experiment had on the data you collected and, therefore, how confident you are that your conclusion is correct.

- Don't forget to use the **secondary data** together with your own data when making your final conclusion. And don't forget to use your scientific knowledge to explain why you have come to this conclusion.

Of the three parts of the Controlled Assessment, Part C is probably the most challenging – mostly because you're doing it by yourself. Remember that you can practise the skills needed to succeed in this part of the Controlled Assessment – use your textbook wisely, as it will have lots of advice on tackling the different aspects of the practical work.

And don't forget that you can have more than one go at any part of the Controlled Assessment – it's your best mark for each part of the task that counts!

Biology 1
Influences on life

It might appear to be a slightly odd living arrangement but this shrimp and goby fish live together. And they both benefit from the deal. The shrimp digs and maintains their burrow. The shrimp has poor eyesight and so relies on the goby to warn it of danger. The goby slaps the shrimp with its tail fin if danger approaches and if the fish goes to hide back in the burrow, the shrimp will follow. The shrimp can also communicate with the goby by touching it with its antennae.

In this unit you will learn about other relationships like this, and about communication systems, both between organisms and inside single organisms. You will also find out how scientists think that life on Earth has developed over time, and how scientists use a classification system for organisms to develop ways in which the rich diversity of life on Earth can be protected for future generations.

How are organisms classified?

Natural sea sponges can be used for cleaning and many people assume that they are plants. Sponges are actually animals and the 'sponge' that we use is a very soft skeleton.

A *Sea sponges grow on the seabed, attached to rocks.*

Until the mid-1700s most scientists thought that sponges were plants. This was due to the Greek thinker Aristotle (384 BCE to 322 BCE) who was one of the first people to sort organisms into groups based on their **characteristics** (what they look like). This process is called **classification**. For Aristotle, organisms that did not move were plants.

> **1** Why did Aristotle think that sponges were plants?
>
> **2** What is classification?

Today, scientists look at all the characteristics of organisms to decide how to group them. Organisms that have many characteristics in common are grouped together as a **species**. Species that share many characteristics are grouped as a **genus** (plural genera). Genera that share many characteristics are grouped as a **family**, and so on up to the level of **kingdom**. Figure B shows the relationships between the major classification groups for the Kingdom Animalia, and shows how the African lion is classified within these groups.

Results Plus
Watch Out!

All cells have cell membranes but they don't all have cell walls. Plant cells also have cellulose cell walls.

> **3 a** To which class does the African lion belong?
> **b** Name one other class in the animal kingdom.
>
> **4** At which level of classification is the African lion grouped with the polar bear?

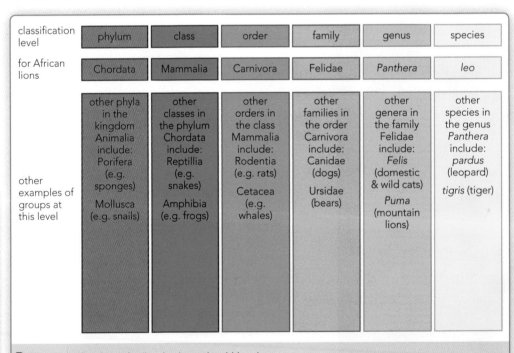

classification level	phylum	class	order	family	genus	species
for African lions	Chordata	Mammalia	Carnivora	Felidae	*Panthera*	*leo*
other examples of groups at this level	other phyla in the kingdom Animalia include: Porifera (e.g. sponges) Mollusca (e.g. snails)	other classes in the phylum Chordata include: Reptillia (e.g. snakes) Amphibia (e.g. frogs)	other orders in the class Mammalia include: Rodentia (e.g. rats) Cetacea (e.g. whales)	other families in the order Carnivora include: Canidae (dogs) Ursidae (bears)	other genera in the family Felidae include: *Felis* (domestic & wild cats) *Puma* (mountain lions)	other species in the genus *Panthera* include: *pardus* (leopard) *tigris* (tiger)

B *The classification of a lion in the animal kingdom*

The five kingdoms of organisms

Organisms may be classified into one of five kingdoms, according to some very basic characteristics. For example, some organisms are **unicellular** (single cells) while others are **multicellular** (made of many cells). Many organisms have cells that contain a **nucleus**, like our cells, but **bacteria** are simple cells with no nucleus. Another basic characteristic is the way that the organism gets its food. Plants make their food **autotrophically** using **photosynthesis**, while animals feed **heterotrophically** by eating and digesting other organisms. Fungi also get their food by digesting other organisms, but they do this outside the body and are said to feed **saprophytically**.

Kingdom	Main characteristics
Animalia	multicellular; heterotrophic feeders so no chlorophyll; no cell walls; complex cell structure with nucleus
Plantae	multicellular; autotrophic feeders using chlorophyll; cell walls made of cellulose; complex cell structure with nucleus
Fungi	multicellular; cell walls not made of cellulose; saprophytic feeders so no chlorophyll; complex cell structure with nucleus
Protoctista	mostly unicellular (a few are multicellular); complex cell structure with nucleus
Prokaryotae	unicellular; simple cell structure with no nucleus

C *The main characteristics of organisms in the five kingdoms.*

There is no kingdom for **viruses** because most scientists do not think of them as being alive. When a virus particle enters a living cell, it changes the way the cell works and causes it to make copies of the virus. However, the actual virus particle does not show other life processes, such as growth or feeding, like other organisms.

7 A dictionary publisher of a children's dictionary needs definitions of the words 'plant' and 'animal'. Write definitions for the dictionary.

Skills spotlight

In a successful argument, evidence is presented for and against an idea and then is used to reach a decision. Fungi used to be classified in the Kingdom Plantae. Construct an argument for classifying fungi as plants or as a separate kingdom.

5 Seaweeds do not have cellulose cell walls but do photosynthesise.
a Suggest why seaweeds were once thought to be plants.
b Give one reason why seaweeds are no longer classified as plants.

6 Viruses do not have a nucleus and nor do bacteria. So why are viruses not classified as Prokaryotae with bacteria?

D *A virus particle*

Learning Outcomes

1.1 Demonstrate an understanding of how biologists classify organisms according to how closely they are related to one another including **a** Species **b** Genus **c** Family **d** Order **e** Class **f** Phylum **g** The Five Kingdoms

1.2 Describe the main characteristics of the five kingdoms including **a** Animalia **b** Plantae **c** Fungi **d** Protoctista **e** Prokaryotae

1.3 Explain why scientists do not classify viruses in any of the five kingdoms and regard them as non-living

HSW **11** Present information, develop an argument and draw a conclusion, using scientific, technical and mathematical language, and ICT tools

Vertebrates and invertebrates

 How is the animal kingdom subdivided?

English settlers in Australia discovered duck-billed platypuses in 1797. Dead specimens were sent back to England but scientists thought they were a practical joke! When the platypus was confirmed as being real, scientists could not agree how to classify a poisonous, furry animal with a duck's beak, a beaver's tail, an otter's feet and a lizard's skeleton shape, that lays leathery eggs.

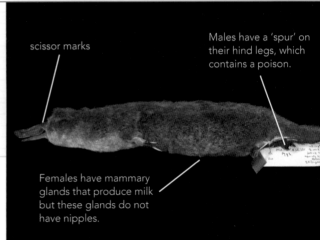

scissor marks

Males have a 'spur' on their hind legs, which contains a poison.

Females have mammary glands that produce milk but these glands do not have nipples.

A *This duck-billed platypus was sent to the Natural History Museum at the end of the 18th century. The scientist who examined it tried to remove the bill with a pair of scissors because he thought it was a hoax.*

The one thing that scientists could agree on was that platypuses were **vertebrates**. Vertebrates are animals that have a backbone (a series of small bones called **vertebrae**). All vertebrates belong to the phylum **Chordata** because they have a supporting rod that runs the length of their body.

1 a What is a backbone?
b What is its purpose?

2 Explain why the duck-billed platypus is a chordate.

3 State one feature that members of the following groups of animals all have in common:
a chordates
b vertebrates.

4 Describe three ways in which different vertebrates obtain oxygen from the environment.

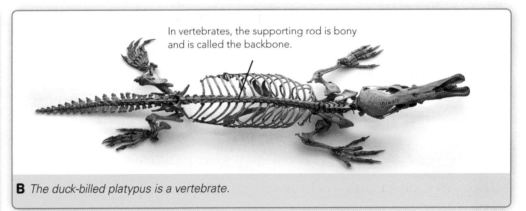

In vertebrates, the supporting rod is bony and is called the backbone.

B *The duck-billed platypus is a vertebrate.*

Animals that do not have a backbone are called **invertebrates**. The vertebrate and invertebrate groups are both divided into smaller groups.

One characteristic we can use to group vertebrates is the way they absorb oxygen for respiration. **Fish** have gills to take oxygen from water. Young **amphibians** also have gills, but adult amphibians usually have lungs and can also absorb oxygen through their moist skin. The other groups of vertebrates (**mammals**, **reptiles** and **birds**) have lungs.

We can also divide vertebrates into groups using reproduction characteristics. Some vertebrates reproduce using **external fertilisation**. This is where the adult female releases her eggs into the water where they are fertilised by sperm released by an adult male. Other vertebrates reproduce by placing the sperm inside the female so that **internal fertilisation** takes place inside her body. Many vertebrates in this group then lay eggs (they are **oviparous**), but mammals give birth to live young (they are **viviparous**).

Another characteristic we can use is whether the vertebrates are **homeotherms**. This means that they keep their body temperature more constant, and often warmer, than their surroundings by releasing heat from reactions in their body. Other vertebrates are **poikilotherms**, which means that their body temperature varies with the temperature of their surroundings.

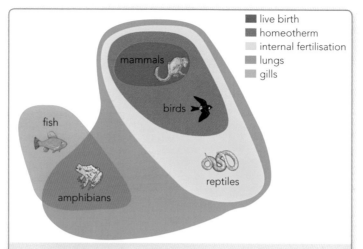

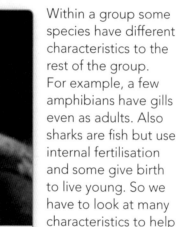

- ■ live birth
- ■ homeotherm
- ■ internal fertilisation
- ■ lungs
- ■ gills

C We can use characteristics to show relationships between vertebrate groups.

D The amphibian axolotl keeps its gills even when it is an adult.

Within a group some species have different characteristics to the rest of the group. For example, a few amphibians have gills even as adults. Also sharks are fish but use internal fertilisation and some give birth to live young. So we have to look at many characteristics to help decide in which group to place an organism.

Skills spotlight

We classify things into groups to help make sense of a large amount of different data. We can then use these groups to develop ideas and theories. Figure E is a Venn diagram and is used to show the relationships between different groups. Draw a Venn diagram to show the relationships between animals, vertebrates and invertebrates.

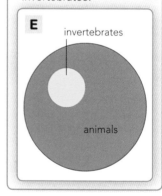

E
invertebrates
animals

5 **a** How are reptiles and birds similar? **b** How are they different?

6 Which characteristic separates birds and mammals from the other vertebrates?

7 Construct an argument for classifying an axolotl as a fish.

8 Explain why it was difficult to classify the platypus as a mammal, and suggest how scientists finally made that decision.

Learning Outcomes

1.4 Describe the main characteristics of the phylum chordata as animals with a supporting rod running the length of the body, an example of this being the backbone in vertebrates.

1.5 Explain how scientists place vertebrates into groups based on **a** oxygen absorption methods – lungs, gills and skin **b** reproduction – internal or external fertilisation, oviparous or viviparous **c** thermoregulation – homeotherms and poikilotherms

1.6 Demonstrate an understanding of the problems associated with assigning vertebrates to a specific group based on their anatomy and reproduction methods and why many vertebrates are difficult to classify

HSW **2** Describe how data are used by scientists to provide evidence that increases our scientific understanding

Species

 Why can classifying species be difficult?

In 2003, there was surprise at a Japanese safari park when a female donkey gave birth to a foal with stripes! Its father had been one of the park's zebras.

A The 'zedonk' hybrid foal with its mother.

1 American robins and British robins cannot interbreed. Suggest why not.

2 What is a hybrid?

3 Why can't a zedonk be classified as a species?

If you look at Figure B on page 28, you will see that the organisms in each box get more and more closely related – they have more in common with each other. The two boxes in the last column show individual **species**. A species is defined as a group of organisms that can **interbreed** (reproduce with one another) to produce offspring that are **fertile** (able to reproduce).

However, this definition of a species is not always clear-cut because two closely related species can often breed and produce **hybrids**. Hybrids are neither one species nor the other, and show characteristics of both parents. Hybrids are usually infertile but not always. Being infertile means that they are unlikely to have offspring of their own and pass on their mix of characteristics.

A further problem with our definition of a species is that some organisms do not always need to interbreed to produce offspring. For example, many plants and fungi can produce new individuals from parts of adult organisms. Also bacteria and many protoctists may reproduce by splitting in half. This means each new individual has just one parent. If we never see interbreeding in an organism, we cannot test whether or not two individuals are the same species.

Difficulties with classification

Mallard ducks can hybridise with other closely related species to produce fertile hybrid offspring. These offspring then breed with other hybrids or mallard ducks or closely related ducks. This produces ducks with a continuous range of characteristics, rather than separate species.

Neighbouring populations of the same species may have slightly different characteristics but still interbreed. Sometimes there is a chain of different populations that can all breed with their neighbouring populations but the two populations at either end of the chain cannot interbreed. The chain often forms a ring shape and so these organisms are called **ring species**.

It is hard to divide ring species into separate species. The gulls in Figure B are traditionally considered to be two species – *Larus fuscus* (the lesser black-backed gull) and *Larus argentatus* (the European herring gull).

Skills spotlight

Scientists tell each other about their work by writing papers, which are published in journals (a type of magazine). Papers are usually peer-reviewed, which means that other scientists read the papers and say whether their scientific evidence is good enough to be published. State one advantage and one disadvantage of peer-reviewing.

However, classifying the gulls between these two is difficult because there is a gradual change of characteristics between the two species.

Naming a species

An organism's scientific name has two Latin words (genus and species). This is called the **binomial system**. Organisms that share the first word in their name (e.g. *Equus*, *Larus*) are closely related. Scientists can tell from the name whether two organisms are closely related.

The binomial system is also useful because organisms with the same common names may actually be different species. For example, a robin in America is not the same bird as a robin in the UK. The system is agreed by scientists all over the world to allow them to communicate clearly, whatever their language.

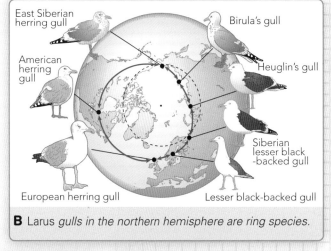

B *Larus gulls in the northern hemisphere are ring species.*

C *An American robin,* Turdus migratorius

D *A UK robin,* Erithacus rubecula

H 4 Suggest two gulls in Figure B that cannot interbreed.

H 5 Why might it be difficult to identify the species of a duck on a local pond?

H 6 Why is it hard to divide populations that form ring species into separate species?

H 7 Why is the binomial system so useful?

8 Explain why it is difficult to clearly define what we mean by the term species.

ResultsPlus
Watch Out!

The word species is well known, but you need to remember that in the binomial system, genus and species are used together and written in italics.

Learning Outcomes

1.7 Discuss why the definition of a species as organisms that produce fertile offspring may have limitations: some organisms do not always reproduce sexually and some hybrids are fertile

H 1.8 Explain why binomial classification is needed to identify and study species

1.9 Explain how accurate classification may be complicated by: **H** *b* hybridisation in ducks **H** *c* ring species

1.19 Explain the role of the scientific community in validating new evidence, including the use of: *a* scientific journals *b* the peer review process *c* scientific conferences

HSW 14 Describe how scientists share data and discuss new ideas, and how over time this process helps to reduce uncertainties and revise scientific theories

What is variation?

In 2009, a BBC camera crew discovered a new species of giant rat, deep in the jungles of Papua New Guinea. It was 82 cm long and not afraid of humans. While scientists are agreeing its scientific name, it is being called the Bosavi woolly rat (after the place in which it was found).

A *A Bosavi woolly rat*

1 a What is variation?
b What variation is there between the rats in Figure B?

Differences in characteristics are called **variation**. There is variation within a species but much more variation between different species. This variation within a species can make classification difficult.

B *These rats are all of the same species but show great variation. Great variation within a species can be confusing and make classification difficult.*

Scientists must make sure that any 'new' organism is not just a hybrid or due to variation in a known species. They do this by finding more than one of the organisms.

From the Bosavi woolly rats that have been found, scientists can tell that they are closely related to other 'hairy rats' because they share many characteristics. However, Bosavi woolly rats have sufficient variations in those characteristics to show that they are a new species.

Keys

To identify different species, you can use a **key**. Figure C shows a key for rats.

2 a What evidence is there that the woolly rat is a different species to the rats in Figure B?
b How might further evidence be collected to show that they are different species?

3 Why do scientists need to find more than one example of a 'new' organism?

4 Use the key to identify which rat species is shown in Figure B.

Statement:	Next step:
1 The animal has a thick, bushy tail.	squirrel
The animal has a tail that is covered with layers of skin, which look like scales.	**go to 2**
2 It has a band of hair that is longer than the rest of the fur.	crested rat
Its fur is all the same length.	**go to 3**
3 It has a short tail.	groove-toothed rat
It has a long tail.	**go to 4**
4 Its tail is shorter than its body.	brown rat (*Rattus norvegicus*)
Its tail is longer than its body.	black rat (*Rattus rattus*)

C *To use the key, pick the correct statement from the first box and follow the 'next step' instructions. Carry on doing this until you reach a name.*

The importance of classification

Accurate classification using the binomial system allows biologists to:
- easily identify existing species and new species
- see how organisms are related
- identify areas of greater and lesser **biodiversity**.

Biodiversity is a measure of the total number of different species in an area. To count the species you need to be able to identify them, which can be tricky if species are very similar to one another. The more accurate your classification system, the easier identification becomes.

Biodiversity is important because we obtain many products from living things (e.g. foods, medicines). The more species there are, the more choices we will have, both now and in the future. Biodiverse areas are also much better at recovering from natural disasters (e.g. floods) than less diverse areas.

Many biologists think that areas of greater biodiversity ('biodiversity hotspots') are the ones that need the most time and money spent on trying to conserve them because this will result in a greater number of species being conserved.

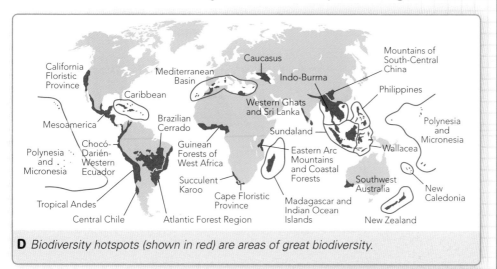

D Biodiversity hotspots (shown in red) are areas of great biodiversity.

ResultsPlus
Watch Out!

It is sometimes difficult to decide between two statements in a key. Be prepared to go back and think again if a key doesn't appear to be working.

Skills spotlight

The paragraph at the top of page 34 contains secondary data because the person using it is not the person who collected the data. The information comes from a TV programme. The people who wrote the TV programme used primary data because they found the rats. Suggest one advantage and one disadvantage of using secondary data.

H 5 How does classification make it easier to measure biodiversity?

H 6 Why is it important to protect biodiversity hotspots?

7 Design a key to identify some farm animals. Try to include animals from different species and from the same species.

Learning Outcomes

H **1.8** Explain why binomial classification is needed to identify, study and conserve species, and can be used to target conservation efforts

1.9 Explain how accurate classification may be complicated by a variation within a species

1.10 Construct and use keys to show how species can be identified

HSW **1** Explain how scientific data is collected and analysed

How tall can people grow?

B1.5 Variation

Who might wear size 28 shoes?

When you measure the variation in a certain characteristic you find a range of values, with some values more common than others. For example, adult humans range in height from 73 cm to 246.5 cm but most are 150 to 190 cm.

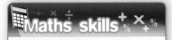

Maths skills

If you want to find the average of a set of quantitative data you can calculate the mean, where

$$\text{mean} = \frac{\text{sum of values}}{\text{number of values}}$$

E.g. If you have five pea pods of length 81 mm, 58 mm, 95 mm, 64 mm and 122 mm, and you want to find the average lenth of the pea pods you can calculate:

mean pod length =

$$\frac{81 + 58 + 95 + 64 + 122}{5}$$

$$= \frac{420}{5}$$

$$= 84 \text{ mm}$$

The range in human heights is caused by both inherited and environmental variation. It is also an example of **continuous variation** – the values can be any number within a certain range.

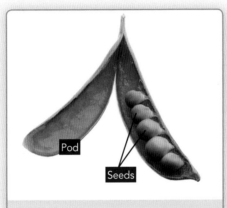

B Many plants have fruit called pods, inside which there are seeds.

A In January 2010, the world's tallest and shortest men met. At the time, Sultan Kosen had size 28 shoes.

Sultan Kosen

He Pingping

Shoe size is an example of **discontinuous variation**. A characteristic with discontinuous variation only has a fixed set of values.

In this case, there is a relationship (link) between foot length and shoe size; people with longer feet wear bigger shoes.

Your task

You are going to plan an investigation that will allow you to find out if there is a link between the length of a pea pod and the number of seeds inside. You are going to test the hypothesis that longer pods contain more seeds. Your teacher will provide you with some materials to help you organise this task.

Learning Outcomes

1.13 Describe variation as continuous or discontinuous

1.14 Investigate the variations within a species to illustrate continuous variation and discontinuous variation

Build Better Answers

When completing an investigation like this, one of the skills you will be assessed on is your ability to assess the *quality of evidence*. There are 4 marks available for this skill. Here are two student extracts that focus on this skill. Other skills that you need for the assessment are dealt with in other lessons.

Student extract 1 — A basic response for this skill

Make it clear which results you think are anomalous. One way to do this is to identify them.

Not all results contain anomalies and you should point this out as well.

> My results show three anomalous results. I have ringed these in my results table. These are anomalous because they are very different from the other results. My secondary evidence has no anomalous results in it.

Look for anomalous results in the secondary evidence as well as your own results.

Student extract 2 — A good response for this skill

Explain how you have dealt with any anomalies when processing your results.

> My results show three anomalous results. I have ringed these in my results table. I have then not included these results when calculating the means for pod length. I have left them out because they do not fit the pattern of the other results. All of the other results seem to fit a regular pattern so I shall use all of these. There are no anomalies in the secondary evidence.

This is not a full answer because you need to explain why you decided to exclude the results.

You should point out when there are no anomalies but you should also explain that this means you will use all your data when processing your results.

ResultsPlus

To access 2 marks
- Comment on the quality of your primary and secondary evidence
- Deal with any anomalies appropriately
- Say if you do not think there are any anomalies in your evidence

To access 4 marks
- Take account of any anomalies in your primary and secondary evidence
- Explain any adjustments you need to make to your evidence
- If you do not think there are anomalies, explain this and say that you are using all your evidence

What causes variation?

Shark tissues contain TMAO, a substance that is poisonous to humans in high concentrations. Greenland sharks, found in the Arctic Ocean, contain much higher amounts of TMAO than other sharks because it is a natural antifreeze. This makes them poisonous for humans to eat.

A *Hákarl, made from rotten shark, is a food speciality in Iceland. Burying the flesh in the ground for a few months gets rid of the TMAO poison.*

All organisms are **adapted** to their surroundings – they have variations in their characteristics that allow them to survive in their **habitats**.

Organisms from polar regions, like Greenland sharks and polar bears, are adapted to the cold. Organisms living near deep-sea **hydrothermal vents** cope with the opposite problem. Hot fluids, at more than 350°C, come out of these vents and cool quickly. The organisms living here must cope with big temperature changes, complete darkness and huge pressures.

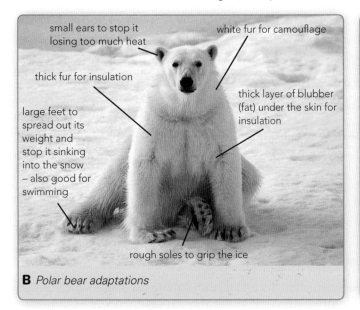

small ears to stop it losing too much heat

white fur for camouflage

thick fur for insulation

thick layer of blubber (fat) under the skin for insulation

large feet to spread out its weight and stop it sinking into the snow – also good for swimming

rough soles to grip the ice

B *Polar bear adaptations*

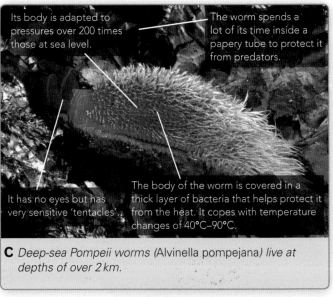

Its body is adapted to pressures over 200 times those at sea level.

The worm spends a lot of its time inside a papery tube to protect it from predators.

It has no eyes but has very sensitive 'tentacles'.

The body of the worm is covered in a thick layer of bacteria that helps protect it from the heat. It copes with temperature changes of 40°C–90°C.

C *Deep-sea Pompeii worms (Alvinella pompejana) live at depths of over 2 km.*

?

1 Which of the polar bear's adaptations help it survive in its habitat?

2 Why doesn't the Pompeii worm need eyes?

Different types of variation

Although all polar bears have thick fur, different individuals may have different thicknesses of fur. Most characteristics show variation between individuals of the same species. Some characteristics show discontinuous variation, such as blood group and gender in humans. An example of a graph which shows the discontinuous variation of human blood groups can be seen in Figure D. Discontinuous variation in characteristics is usually caused by the instructions within your cells and is called **genetic variation**.

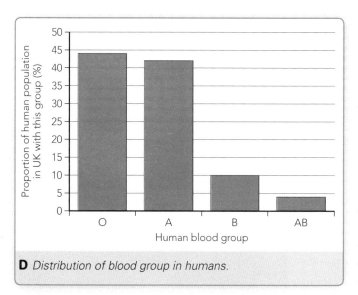

D *Distribution of blood group in humans.*

If a characteristic shows continuous variation, and we measure a large enough number of individuals, when we plot the results we often get a graph with a particular shape. This shape is called a **normal distribution curve**. It shows that most individuals measure within the middle part of the range in variation, and there are fewer individuals with measurements at the extremes of the range.

Characteristics that show continuous variation are often controlled by both genes and the environment. You may inherit a tendency for being tall from your parents. However, if you don't eat healthily in childhood, or you have an accident that damages your bones, you may never grow tall. Your shortness will have been an **acquired characteristic**. Acquired characteristics are caused by the environment and so cause **environmental variation** between organisms.

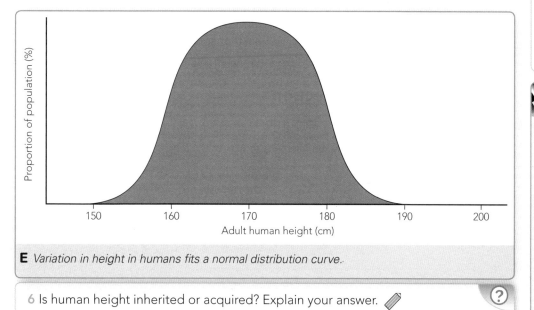

E *Variation in height in humans fits a normal distribution curve.*

6 Is human height inherited or acquired? Explain your answer.

3 Explain why human blood group and gender are examples of discontinuous variation.

ResultsPlus
Watch Out!

Very few characteristics show only inherited variation, or only variation caused by the environment. Most characteristics are produced by an interaction of inheritance and environment.

4 What is the range of human height shown in Figure E?

5 Describe the graph in Figure E in your own words.

Skills spotlight

Some data that we collect are qualitative, meaning that they are descriptions. For example, qualitatively most people in the UK have blood group O. Other data are quantitative, meaning that they have measured values, such as shoe size. Suggest how you could collect both qualitative and quantitative data on eye colour.

Learning Outcomes

1.11 Explain how organisms are adapted to their environment and how some organisms have characteristics that enable them to survive in extreme environments, including deep-sea hydrothermal vents and polar regions

1.15 Interpret information on variation using normal distribution curves

1.16 Demonstrate an understanding of the causes of variation, including:
 a genetic variation – different characteristics as a result of mutation or reproduction b environmental variation – different characteristics caused by an organism's environment (acquired characteristics)

HSW *10* Use both qualitative and quantitative approaches to collecting data.

What is Darwin's theory of evolution?

Most crocodiles live in or near water, but there are crocodiles that survive in the Sahara Desert. They survive without food for the many months when it is too hot and dry, staying below ground in a kind of hibernation. When the rains eventually arrive they come to the surface to hunt.

A

Organisms generally produce far more offspring than the environment can support. Most will die before they reach adulthood because there are not enough resources (e.g. food and space) for them all.

1 a Why is there competition between individuals in a population?
b What effect will competition have on the population?

2 Explain what is meant by 'survival of the fittest'.

The offspring will show variation in their characteristics. Some variations will be better adapted to the environment than others. The limited resources will cause **competition** between individuals. Individuals that happen to have variations that are better adapted will be more likely to survive and others will die. We call this **'survival of the fittest'** or **natural selection**. Individuals that survive may breed and so may pass their variations on to their offspring.

Evolution by natural selection

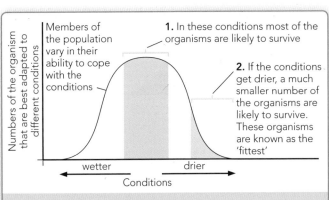

Numbers of the organism that are best adapted to different conditions

Members of the population vary in their ability to cope with the conditions

1. In these conditions most of the organisms are likely to survive

2. If the conditions get drier, a much smaller number of the organisms are likely to survive. These organisms are known as the 'fittest'

wetter ← Conditions → drier

B *If conditions change, the range of variation favoured by natural selection could change.*

Evolution means 'a gradual change over time'. In the early 1800s, some scientists realised that species can change over time, but theories of why this happened didn't explain all the facts. Charles Darwin (1809–1882) drew together several ideas to produce a new theory that was better at explaining the facts.

Darwin knew that there was competition between individuals. He also knew that pigeon breeders selected birds with certain characteristics to breed, which meant that those characteristics were inherited by the offspring. Darwin realised that if the environment changes, then different variations may be better suited to the new conditions. Individuals with those variations will be more likely to survive and pass their characteristics on to offspring. So the range of variation in characteristics of the population will gradually change over generations, which is evolution. If the environment changes too rapidly, and no individuals have adaptations that help them survive, they will all die out and the species may become **extinct**.

Darwin published his theory in his book *On the Origin of Species by Means of Natural Selection* in 1859. Some scientists didn't like the theory because it didn't support what they believed. The theory was discussed and challenged at scientific conferences by scientists studying similar areas of science.

New evidence for Darwin's theory

One of the problems with Darwin's theory when he published it was that there was little evidence for it. Evolution takes time to observe.

Warfarin is a chemical that was used to poison rats in the 1940s and 1950s. When it was first used most rats that ate it died within a few days. But within 10 years, most rats were **resistant** to warfarin and not affected by it. As a result of variation, there had always been a few rats that were resistant, but at the start nobody realised this. As the poison killed the non-resistant rats, the only ones left to breed were resistant. Scientists are also finding new evidence for Darwin's theory from the genetic information in different species. Darwin could infer relationships between species from the observable characteristics but now we are able to confirm this using genetic evidence.

Speciation

Darwin started thinking about evolution after noticing differences between mockingbirds from different Galapagos Islands. He realised they were all very closely related, but each island had its own species. His theory helped him explain this observation. Darwin guessed that originally individuals from one species of mockingbird had reached the islands from South America. The environmental conditions varied between islands, so on each island different adaptations would have been more successful. So each island population evolved in a different way. Over time, the individuals on each island became so different that they couldn't interbreed with birds from another island. They had become new species, a process called **speciation**.

C *Española Island mockingbird*

D *Santiago Island mockingbird*

ResultsPlus
Watch Out!

There are different theories of evolution, not just Darwin's, so refer to *Darwin's* theory of evolution by natural selection.

3 Explain how a species may become extinct.

4 Why was Darwin's theory of evolution better than earlier theories?

5 Explain how warfarin resistance evolved in rat populations.

6 State one difference between the mockingbirds in Figures C and D.

7 Ground finches have large, powerful beaks to crush seeds. A closely related species has a narrow beak for probing in small holes for insect larvae. Suggest how this species could have evolved from the seed-eating species.

Learning Outcomes

1.12 Demonstrate an understanding of Darwin's theory of evolution by natural selection including *a* variation *b* over-production *c* struggle for existence *d* survival *e* advantageous characteristics inherited *f* gradual change

H *1.17* Demonstrate an understanding of how speciation occurs as a result of geographic isolation

1.18 Explain how new evidence from DNA research and the emergence of resistant organisms support Darwin's theory

HSW *2* Describe the importance of creative thought in the development of hypotheses and themes.

Where is the information for variation stored?

Nearly 25 percent of all mammal species are bats! Scientists think that bats evolved from mouse-like animals about 50 million years ago. They think that a group of these animals suddenly got very long 'fingers', which they started to use as wings. The sudden change was caused by a mistake in the 'instructions' inside cells.

A The heart-nosed bat can hear an insect walking from two metres away.

Three of the main parts of most cells are the **cell membrane**, the **cytoplasm** and the **nucleus**. Inside the nucleus there are long strands of a substance called **DNA**. Each strand forms a structure called a **chromosome**.

1 a Draw an animal cell. Label the cell membrane, cytoplasm and nucleus.
b In which part of the cell are chromosomes found?
c What are chromosomes made of?

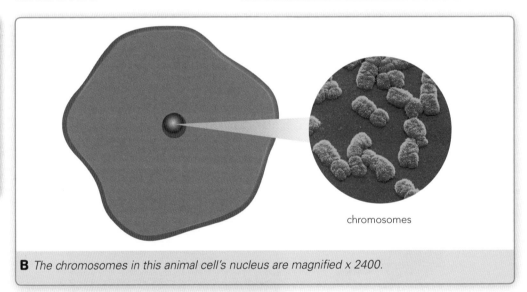

chromosomes

B The chromosomes in this animal cell's nucleus are magnified x 2400.

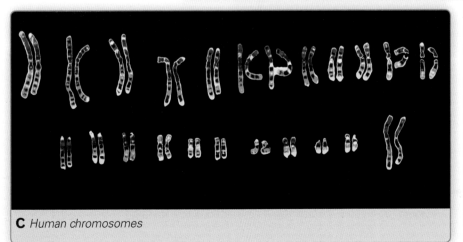

C Human chromosomes

A nucleus contains different chromosomes. There are usually two copies of each type of chromosome.

Chromosomes are divided up into **genes**. Each chromosome carries a large number of genes and each gene does a particular job. For example, many genes control variations in our characteristics – what we look like (e.g. eye colour, face shape). Variation caused by genes is **inherited variation** because we inherit our genes from our parents.

You can think of chromosomes as a set of books. Each book (chromosome) contains a set of sentences giving instructions (genes). All of the books together contain all of the instructions needed to produce a certain organism.

Alleles

Some genes for the same characteristic (e.g. eye colour) may contain slightly different instructions to create variations (e.g. brown, blue). Different forms of the same gene are called **alleles**.

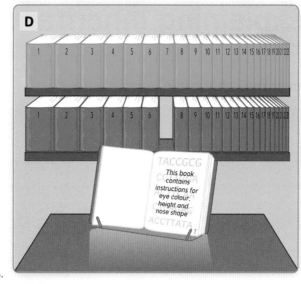

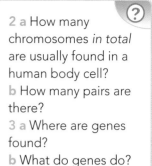

2 a How many chromosomes *in total* are usually found in a human body cell?
b How many pairs are there?
3 a Where are genes found?
b What do genes do?

ResultsPlus
Watch Out!

In exam questions about the differences between the same characteristic, write about alleles (and not genes).

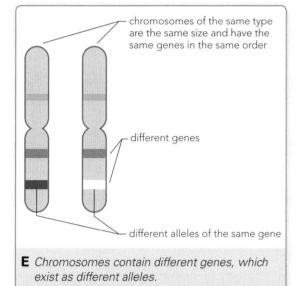

- chromosomes of the same type are the same size and have the same genes in the same order
- different genes
- different alleles of the same gene

E *Chromosomes contain different genes, which exist as different alleles.*

Since there are two copies of every chromosome in a body cell nucleus, there are two copies of every gene. Each copy of a gene may be a different allele.

Different organisms have different numbers of chromosomes. Human body cell nuclei contain 23 pairs of chromosomes, which contain about 23 000 different genes in total. There are many alleles for each gene, so it is easy to see why each of us can inherit a different set of alleles from our parents. Each different set of alleles gives each of us slightly different characteristics.

Skills spotlight

Scientists are busy finding out what all of our alleles do. Imagine that an allele is discovered that causes a disease in older people. Someone suggests that all babies are tested for this allele. Suggest one advantage and one disadvantage of this testing.

4 What are alleles?

5 Where, inside a cell, did the mistake occur that caused some mouse-like animals to get long fingers?

6 How does the idea of alleles help us to explain why we all look different?

7 How would you add to the book analogy above to include alleles?

8 Use your knowledge of cells and genes to explain how a scientist would find out whether blood at a crime scene belongs to the victim or to a potential suspect.

Learning Outcomes

1.20 Describe the structure of the nucleus of the cell as containing chromosomes, on which genes are located

1.21 Demonstrate an understanding that genes exist in alternative forms called alleles which give rise to differences in inherited characteristics

HSW **13** Explain how and why decisions that raise ethical issues about uses of science and technology are made

If you breed a pea that has purple flowers with one that has white flowers, what colour flowers will the offspring have?

Explaining inheritance

How can we predict some inherited characteristics?

An Austrian monk first put forward the idea of genes and alleles in 1865. His name was Gregor Mendel. Chromosomes were not known about at the time, so his ideas were based on observations of how pea plants inherited discontinuous characteristics.

A *Gregor Mendel (1822–1884) grew nearly 30 000 pea plants in his investigations.*

1 How many chromosomes are in a normal human sperm cell? ⑦

2 Why does a pea pollen grain contain only one flower colour allele? ⑦

Plants and animals produce **gametes (sex cells)**. The male gametes are **sperm cells** in animals and **pollen grains** in plants. The female gametes are **egg cells** in plants and animals. Gametes are different to most body cells because they only have one copy of each chromosome.

Gametes therefore only have one allele for each gene. In sexual reproduction two gametes fuse together. The new organism that is formed contains two alleles for each gene (one from the male parent and one from the female).

In Figure B, the offspring receives two alleles for flower colour from its parents – one white and one purple. However, only the allele for purple flowers has an effect. It is said to be **dominant**. The white flower allele has no effect if the purple flower allele is also there. This white flower allele is described as **recessive**.

Skills spotlight

The Punnett square shows scientific conventions (standard ways of doing things). Look at Figure D. Explain why the conventions used in Punnett square diagrams are useful.

ResultsPlus
Watch Out!

When writing genotypes, capital letters are written before the small ones (i.e. Rr and not rR).

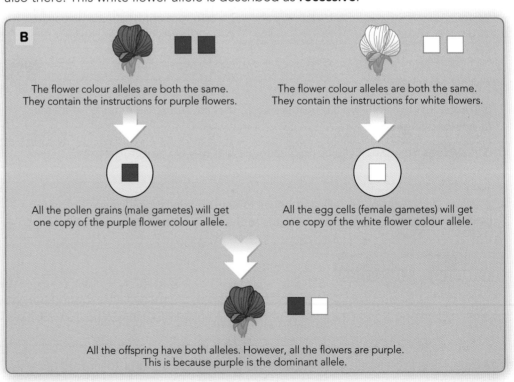

B

The flower colour alleles are both the same. They contain the instructions for purple flowers.

The flower colour alleles are both the same. They contain the instructions for white flowers.

All the pollen grains (male gametes) will get one copy of the purple flower colour allele.

All the egg cells (female gametes) will get one copy of the white flower colour allele.

All the offspring have both alleles. However, all the flowers are purple. This is because purple is the dominant allele.

A recessive characteristic is only seen if *both* alleles are recessive. This can be shown in a **genetic cross diagram**.

A dominant allele is shown by a capital letter (e.g. R for purple). The recessive allele has the lower case version of the *same* letter (e.g. r for *not* purple). The alleles in an organism are its **genotype**. What the organism looks like is its **phenotype**.

If both alleles for a gene in an organism are the same, the organism is **homozygous**. If they are different, it is **heterozygous**. Both parents in Figure C are heterozygous for flower colour.

C *A genetic cross diagram shows the possible combinations of alleles when two organisms breed.*

Punnett squares

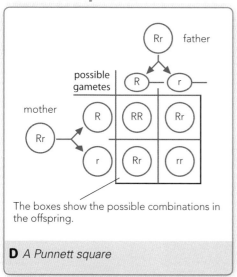

The boxes show the possible combinations in the offspring.

D *A Punnett square*

The possible genotypes produced when two organisms breed can also be shown in a **Punnett square**.

In Figure D there are four boxes of possible genotypes. Three boxes contain genotypes that cause the purple flower phenotype (RR and Rr). From the Punnett square you can work out the **probability** of each phenotype occurring if these plants breed.

$\frac{3}{4}$ = 0.75 probability of purple

or 0.75 × 100 = 75% probability of purple
or 3:1 probability of purple

> 3 When will a recessive allele affect a phenotype?

Maths skills

Probabilities can be written in three ways: as a number between 0 and 1, as a percentage or as a ratio.

- 1 or 100% means something will definitely happen.
- 0.75, 75% or 3:1 means it is quite likely to happen.
- 0 or 0% means it definitely won't happen.

4 The pea plant gene for height has two alleles: T (dominant, causing tall plants) and t (recessive, causing short plants). **a** Draw a Punnett square for breeding a homozygous short plant with a homozygous tall plant. **b** What is the percentage probability of getting a tall phenotype plant?

5 Pea seeds are either yellow or green. Yellow seeds are caused by the dominant allele Y. Show how you would work out the number of the different phenotypes you would get if you crossed two heterozygous pea plants.

Learning Outcomes

1.22 Recall the meaning of, and use appropriately, the terms: dominant, recessive, homozygous, heterozygous, phenotype and genotype

1.23 Analyse and interpret patterns of monohybrid inheritance using a genetic diagram and Punnett squares

1.24 Calculate and analyse outcomes (using probabilities, ratios and percentages) from monohybrid crosses

HSW *11* Present information using scientific conventions and symbols

Suggest a use for plastic blood cells.

What are genetic disorders?

Professor Joseph DeSimone has developed an idea for plastic blood cells. These are tiny plastic sacs that could transport oxygen around the body. This may help people who suffer from some blood diseases.

A *Joseph DeSimone (1964–)*

Skills spotlight

Decisions made by the scientist who invents a new technology can affect how the technology is used. DeSimone applied for a patent on his idea. If you get a patent then other people cannot copy your idea. Why do you think scientists often patent their ideas?

One group of people who may benefit from DeSimone's research are people with **sickle cell disease**. This is a **genetic disorder** – a disease caused by faulty alleles. The sickle cell allele is recessive, so people who have two copies of the allele suffer from the disorder.

People with the disease easily become very tired and short of breath. Sufferers can also have times when their joints are incredibly painful because their red blood cells stick together and block blood vessels. The blocking of blood vessels can sometimes be fatal.

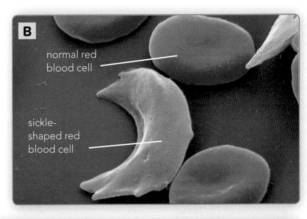

B

normal red blood cell

sickle-shaped red blood cell

Cystic fibrosis

Another genetic disorder caused by a recessive allele is **cystic fibrosis (CF)**. In CF, the lungs get clogged with thick mucus making breathing difficult and leading to infections. Mucus also blocks some of the tubes that carry enzymes to the small intestine to digest food. Lack of enzymes can result in weight loss.

1 Why can't people catch sickle cell disease, like catching a cold?

2 How might a person with sickle cell disease benefit from plastic blood cells?

3 Will someone who is heterozygous for the sickle cell allele have the disease? Explain your answer.

4 Why may CF sufferers lose weight?

5 State *two* symptoms of:
a cystic fibrosis
b sickle cell disease.

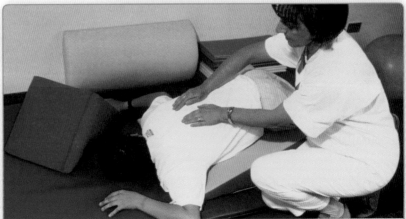

C *Having the back beaten in a special way, followed by coughing, helps to clear the lungs in people with CF.*

Family pedigree charts

Figure D is a **family pedigree chart**, showing how cystic fibrosis is inherited. These charts allow you to see how a genetic disorder is passed on in a family.

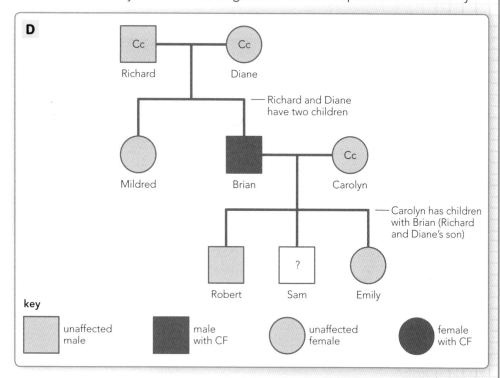

D

— Richard and Diane have two children

— Carolyn has children with Brian (Richard and Diane's son)

key

unaffected male	male with CF	unaffected female	female with CF

ResultsPlus
Watch Out!

When studying a pedigree chart, make sure you read and understand the key first. What exactly do the different shapes mean?

H 6 Look at Figure D.
a What is the recessive allele?
b What is Brian's genotype?

H 7 Predict how likely it is that Sam has CF.

H 8 When Robert and his partner Clare were thinking about having children, their doctor looked at Clare's family pedigree chart as well as Brian's. The doctor advised there was very little risk of them having a child with CF. Suggest what Clare's chart looked like.

Carriers

Doctors can use pedigree charts, and an understanding of how a genetic disorder is inherited, to work out the probability that a person may have inherited the disorder from their parents. This is called **pedigree analysis**. If the risk is high, the doctor may get tests done to see if they have the faulty allele. The doctor will also advise the person on how to stay as healthy as possible.

Doctors also do genetic screening for couples who know that there is a genetic disorder in the family and are worried about passing it on to their children. The doctor will give advice on how to interpret the results and help the couple decide whether to try for a baby or not.

9 CF is a recessive genetic disorder. Explain fully what this means.

1.23 Analyse and interpret patterns of monohybrid inheritance using a genetic diagram and family pedigrees

1.25 Describe the symptoms of the genetic disorders: **a** sickle cell disease **b** cystic fibrosis

H *1.26* Evaluate the outcomes of pedigree analysis when screening for genetic disorders: **a** sickle cell disease **b** cystic fibrosis

HSW *13* Describe the social, economic and environmental effects of decisions about the uses of science and technology

How do we control our internal environment?

In 2005, Dr Stephen Gitelman was puzzled by two babies whose bodies were swollen with water. Working with other doctors, he eventually discovered that rare alleles of a gene called AVPR2 caused the condition. The genetic disorder, now called NSIAD, affects the kidneys.

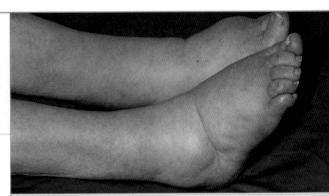

A *Kidney problems can also cause water to collect in parts of the body in adults.*

Water and salt content

The conditions inside the body (the **internal environment**) must remain stable. For instance, it needs enough water for substances to dissolve and for chemical reactions to take place inside the cells. Too much water, however, can cause swellings and high blood pressure. Temperature is also kept constant so that the enzymes that help reactions to occur can work quickly. Keeping the internal environmental stable is called **homeostasis**.

The body loses water in breath and sweat. Sweat is produced by **sweat glands** in the skin (a **gland** is a part of the body that makes substances and then releases them). It also loses water in **urine**, and if the body contains too much water the **kidneys** produce more urine. If the body does not have enough water, the kidneys produce little urine and the brain also responds by making you feel thirsty. The control of water in the body is called **osmoregulation**. The body also controls the concentration of some materials in the body, such as glucose in the blood. **Blood glucose regulation** is important because if glucose concentration gets too high or falls too low, you can become very ill.

Thermoregulation

The body maintains a temperature of 37 °C. This is needed because many chemical reactions in your body work best at 37 °C. The control of body temperature is called **thermoregulation**.

The **hypothalamus** is a small part of the brain that constantly monitors temperature. It receives information from nerve endings in the **dermis** of the skin about the temperature outside the body. It also receives information about the temperature inside the body from the blood. If the body temperature goes below 37 °C, the hypothalamus causes muscles to shiver. This releases heat which warms you up.

The hypothalamus also causes **erector muscles** in the dermis to contract. This causes the body hairs to stand upright. In humans this has little effect, but in other mammals it can trap more air next to the skin as insulation. Oils released from the **sebaceous glands** at the base of the hairs keep the skin lubricated and in good condition. The hypothalamus also reduces blood flow near the skin, keeping the warm blood deeper in so that heat loss to the air is reduced.

B *Water vapour in your breath condenses on cold days.*

?

1 What is your 'internal environment'?

2 What is homeostasis?

3 What is getting thirsty a sign of?

ResultsPlus
Watch Out!

Blood vessels under the surface of the skin widen or narrow. Be careful not to say that they move closer to or away from the surface.

If body temperature goes above 37 °C, the hypothalamus detects this and causes sweating. As sweat evaporates it transfers heat energy from the skin to the surroundings, so the skin cools down. The hypothalamus also increases blood flow nearer to the surface of the skin, making us look pinker. This makes it easier for the blood to lose heat to the air, so we cool down.

4 Why do people shiver when it is cold? ?

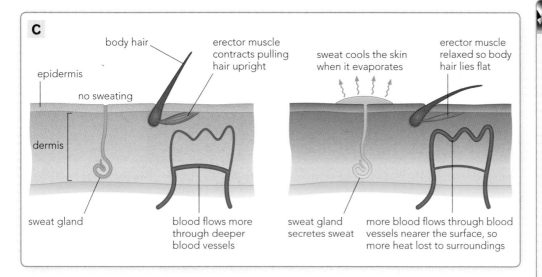

C

body hair
epidermis
no sweating
dermis
sweat gland

erector muscle contracts pulling hair upright
blood flows more through deeper blood vessels

sweat cools the skin when it evaporates

erector muscle relaxed so body hair lies flat
sweat gland secretes sweat
more blood flows through blood vessels nearer the surface, so more heat lost to surroundings

If the body temperature goes above 37 °C, the brain detects this and causes sweating. As sweat evaporates it absorbs heat energy from the skin and so cools it.

H

Skin and body temperature

When it is cold, the hypothalamus reduces blood flow to the skin by narrowing the blood vessels closest to the surface. This is called **vasoconstriction**. The opposite (**vasodilation**) happens when the body is hot and needs to lose heat.

The control of body temperature is an example of **negative feedback**. This means as a change to the body happens in one direction (e.g. increase in heat), mechanisms in the body work to make it change in the opposite direction. This helps keep conditions in the body under control at around the right level.

7 You spend an hour in a greenhouse on a hot day. Describe the changes that happen to your body and explain why they happen. ?

Skills spotlight

Scientists use models to help explain complex processes. A room thermostat controls room temperature by switching a heating system on when the room is below a set temperature and off when the room is above the set temperature. Evaluate how good the model of a thermostat is when used to explain how the body maintains a constant internal temperature.

H 5 Explain vasoconstriction and vasodilation in terms of energy transfer. ?

H 6 Draw a diagram to show how thermoregulation in the human body is a negative feedback mechanism.

Learning Outcomes

2.1 Define homeostasis as the maintenance of a stable internal environment

2.2 Demonstrate an understanding of the homeostatic mechanisms of: **a** thermoregulation and the effect of temperature on enzymes **b** osmoregulation **c** blood glucose regulation

2.3 Explain how thermoregulation takes place, with reference to the function of the skin, including:
   a the role of the dermis – sweat glands, blood vessels, nerve endings, hair, erector muscles and sebaceous glands
   b the role of the hypothalamus – regulating body temperature

H 2.4 Explain how thermoregulation takes place, with reference to:
   a vasoconstriction **b** vasodilation **c** negative feedback

HSW 3 Describe how phenomena are explained using scientific models

Sensitivity

How do animals detect stimuli?

Although the body can usually detect that it is getting cold, sometimes it cannot generate enough heat to maintain the body temperature at 37 °C. The body then gets too cold – a condition called hypothermia. Its symptoms include uncontrollable shivering, tiredness, pain and confusion.

A *The rescue of 21-year-old skier, Cedric Genoud, who was buried under an avalanche for 17 hours. He was found alive but suffering from hypothermia.*

1 What is a stimulus?

2 In which organ are light receptor cells found?

3 Which sense organ do you think contains the greatest number of receptors of different types? Explain your answer.

4 State one way in which the brain alters the way the body works to stop it getting too cold.

Sense organs detect changes, both in and outside your body. Scientists used to think that humans had five senses – touch, sight, hearing, taste and smell. Now we know that we have many more, including heat, cold, pain, balance and changes in position. Anything your body is sensitive to, including changes that you detect in your surroundings, is called a **stimulus**. Sense organs contain **receptor cells**, which detect a stimulus. For example, your skin contains certain receptor cells to detect the stimulus of temperature change.

Receptor cells create electrical signals, called **impulses**, which usually travel to the brain. The brain then processes this information and can send impulses to other organs to alter the way the body works.

ResultsPlus
Watch Out!

Don't confuse neurones with nerves. One nerve contains many neurones.

Electrical impulses travel along cells called **neurones** (or **nerve cells**). This travelling, or transmission, of impulses is called **neurotransmission**. Neurones have a cell body and long extensions that carry the nerve impulse. The **dendron** has many branches at the end called **dendrites** that receive impulses from receptor cells or other neurones. The impulse moves along the dendron, past the cell body to the **axon**. The impulse then passes along the axon to the endings where it passes across to other neurones.

5 Why is it necessary for neurones to link to each other?

6 Draw a flowchart to show how impulses caused by a stimulus reach the brain.

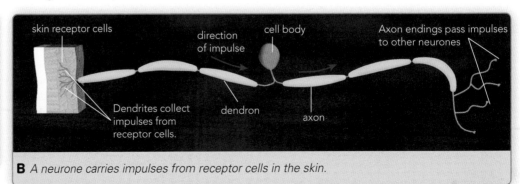

B *A neurone carries impulses from receptor cells in the skin.*

Bundles of neurones are packed together into **nerves**. An organ called the **spinal cord**, which connects to the brain, contains many nerves packed together. The brain and the spinal cord together form the **central nervous system** (**CNS**). The CNS controls your body.

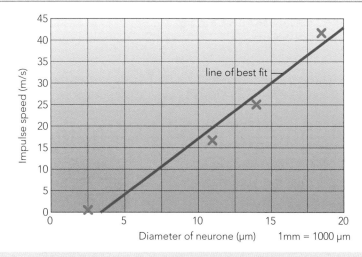

C *A graph showing how the speed of neurotransmission changes with the thickness of the neurone. Source: http://biology.ucsd.edu*

Many investigations involve looking at how variables are linked. Links between the variables are called relationships or correlations. What relationship can you see in Figure C?

Maths skills

Figure C is a **scatter graph**.

Scatter graphs are drawn from pairs of observations and are used to show whether there is any correlation between two variables.

• Positive correlation if one value increases as the other increases.

• Negative correlation if one value increases as the other decreases.

• No correlation if the points are random and widely spread.

The line of best fit is a line that passes as near as possible to the points on the scatter graph.

7 a Which organs are in the CNS?
b Describe the route impulses take from a touch receptor cell in the foot to the brain.

8 You pick up an ice cube. Explain all the different ways in which your brain knows that it is an ice cube.

---

the central nervous system (CNS)

brain

the spinal cord is 43–45 cm long in an adult

some of the thicker nerves outside the CNS

one of the 24 bones (vertebrae) that form the backbone

the sciatic nerve is the thickest in the body – about 1.5 cm in diameter

nerve

blood vessels

bundle of neurones

D *Some bones, the CNS and some of the thicker nerves in a human body*

Learning Outcomes

2.19 Recall that the central nervous system consists of the brain and spinal cord and is linked to sense organs by nerves

2.20 Explain the structure and function of dendrons and axons in the nervous system

2.21 Describe how stimulation of receptors in the sense organs sends electrical impulses along neurones

HSW **11** Draw a conclusion, using scientific and mathematical language

>>>>>>>>>>>>>>>>>>>>>>>>> What do you think an esthesiometer is used for?

B1.13   Skin sensitivity

Where is your skin most sensitive?

You can't tickle yourself because your brain knows when you are going to touch your skin and doesn't react to those sensations as much as if something else touched you. During medical examinations, doctors may put your hand on theirs as they carry out the examination. This stops ticklish feelings because your brain thinks that it's your hand doing the examination – and you can't tickle yourself.

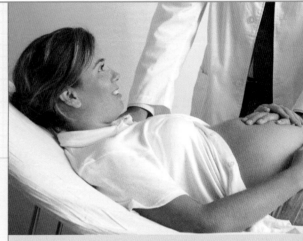

A *You cannot tickle yourself.*

Your skin can detect many different stimuli, including touch, pressure, pain, heat and cold. For each of these stimuli there is a different sort of receptor cell.

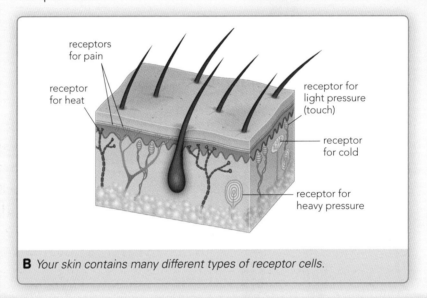

B *Your skin contains many different types of receptor cells.*

Your task

You are going to plan an investigation that will allow you to find out about touch receptors in the skin. You are going to test the hypothesis that the skin on different parts of the arms and hands contains different densities of touch receptors. Your teacher will provide you with some materials to help you organise this task.

Learning Outcomes

2.22 Investigate human responses to external stimuli

When completing an investigation like this, one of the skills you will be assessed on is your ability to *plan a practical so that it tests your hypothesis.* There are 4 marks available for this skill. Here are two student extracts that focus on this skill. Other skills that you need for the assessment are dealt with in other lessons.

Student extract 1 — A basic response for this skill

Think about the range of results you are collecting. Only gathering information from one area of the hand and arm is quite limited but the range of distances is good and tests a lot of the small distances and some large ones too.

> With a partner I am going to decide which area of the arms and hands to test. I will then use my piece of equipment with a gap of 5 mm. I will press this lightly to the skin and they will tell me if they can feel one or two points. I will test one area on the arm and one on the hand. My partner will need to be blindfolded. I will then use different gaps such as 1 mm, 2 mm, 3 mm, 4 mm, 10 mm, 15 mm, 20 mm and 30 mm.

Your method should be in a logical order so that it is easy to follow. It would be better to start with 1 mm and work upwards.

Once again this should be in a logical order – so this important piece of practical information should be early on in the method.

Student extract 2 — A good response for this skill

You also need to explain why you have chosen to take this range of measurements and how they will allow you to test your hypothesis.

> I will use a piece of equipment with two prongs and I will set the prongs to 1mm apart. I will then press either one or two ends to the skin in the area and ask my blindfolded partner how many they can feel. I will test each area three times in different places on the arm and hand. I will then move the prongs apart to 2mm and test again. I will also test 3mm, 4mm, 5mm, 10mm, 20m and 30mm. Using this range will allow me to test where touch receptors are close together (because I am using lots of small measurements) but will also test areas where there are fewer touch receptors (because I am testing some big distances too.) This will be enough to test my hypothesis because I am testing three different areas on the arms and hands and I am testing to see if people can tell the difference between one and two prongs when the gap between them is quite small and quite large.

Make sure you use the right units.

You need to explain how your practical will allow you to test your hypothesis.

ResultsPlus

To access 2 marks
- Provide a logically ordered method
- Choose a range of data or observations that will test the hypothesis

To access 4 marks
You also need to:
- Explain how your method will test the hypothesis
- Explain why you have chosen your range of data or observations

Responding to stimuli

 How do our bodies respond to stimuli?

The device in Figure A allows blind people to see … with their tongues! The image from the camera is sent to a lollipop that contains as many as 600 small electrodes. Each electrode produces pulses of electricity depending on how much light is in that part of the image. By putting the lollipop on the tongue, the user can feel these pulses and build up an idea of basic shapes and movement. This allows some blind people to react and respond to visual stimuli.

A *The BrainPort vision device*

1 Imagine you see a lion and run away.
a Where are the receptor cells that receive the stimulus?
b What effectors carry out the response?
c Suggest another effector that may be triggered by the shock of seeing the lion.

When the brain receives impulses from receptor cells it coordinates a **response**. In the response, impulses are sent to **effectors** and these carry out an action. Effectors include muscles and glands (e.g. sweat glands).

Different neurones

The neurones that receive impulses from receptor cells are **sensory neurones**. An example is shown in Figure B on page 50. These neurones have a long dendron as well as an axon. The neurones that take impulses to effectors are **motor neurones**. These neurones have no dendron, the dendrites are on the cell body. **Relay neurones** are short neurones that are found in the spinal cord, where they link motor and sensory neurones. They also make up the brain.

Watch Out!

Receptors receive stimuli and send impulses along a sensory neurone. Effectors cause an effect.

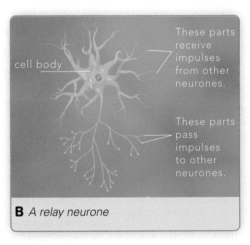
cell body

These parts receive impulses from other neurones.

These parts pass impulses to other neurones.

B *A relay neurone*

2 Do the following carry information to or away from the central nervous system?
a motor neurones
b sensory neurones

Many neurones have a fatty layer surrounding the axon. This is called the **myelin sheath**. The sheath helps to insulate the neurone from surrounding tissue, especially other neurones and also allows impulses to be carried faster.

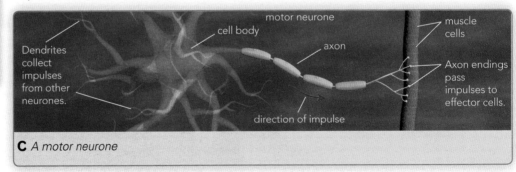
motor neurone
cell body
axon
muscle cells
Dendrites collect impulses from other neurones.
Axon endings pass impulses to effector cells.
direction of impulse

C *A motor neurone*

Synapses

Where one neurone connects with another, there is a small gap called a **synapse**. Impulses are transmitted across the gaps in synapses by chemical substances called **neurotransmitters**. All impulses are slowed down slightly by synapses. However, having synapses is useful because the chemicals are only released from axon endings. This makes sure the impulse can only flow in one direction.

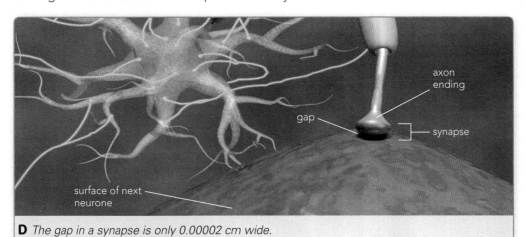

axon ending

gap

synapse

surface of next neurone

D *The gap in a synapse is only 0.00002 cm wide.*

The reflex arc

If you pick up a very hot object you must drop it as soon as possible to stop it burning you, and you don't want to have to think about it first. For these situations **reflexes** are used. Reflex actions are responses that are automatic, extremely quick and protect the body. They use neurone pathways called **reflex arcs** in which a sensory neurone directly controls a motor neurone. Reflex arcs bypass the

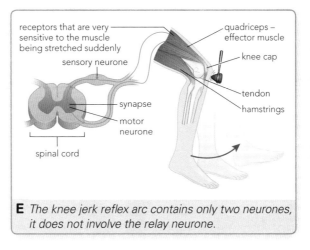

receptors that are very sensitive to the muscle being stretched suddenly

sensory neurone

synapse

motor neurone

spinal cord

quadriceps – effector muscle

knee cap

tendon

hamstrings

E *The knee jerk reflex arc contains only two neurones, it does not involve the relay neurone.*

parts of the brain involved in conscious thought and so are quicker than coordinated responses (responses that involve conscious thought). For example, when the area just below the kneecap is tapped, the knee jerk reflex takes only about 0.05 seconds to occur. This reflex arc contains just two neurones, although some reflex arcs have an 'interneurone' to connect the sensory neurone to the motor neurone.

Other human reflexes include the pupils in your eyes getting smaller in strong light and blinking if something flies towards your face.

Skills spotlight

Botox® injections stop impulses from motor neurones reaching muscle cells. This stops the muscle cells working so that the skin doesn't wrinkle. Think up one question that science can answer about Botox® and one that it cannot answer.

3 Jacob has written: 'An impulse from a neurone travels through a synapse to get to the next neurone.' Explain why this is not correct.

4 Why is the blink reflex useful?

5 Draw a table to compare reflex actions with coordinated responses.

6 Why is it important for reflex arcs to have as few synapses as possible?

7 You kick a football. Describe how this coordinated response occurs. You should mention neurones and the central nervous system in your answer. ✏

Learning Outcomes

2.23 Describe the structure and function of sensory, relay and motor neurones and synapses including:
- **a** the role of the myelin sheath
- **b** the role of neurotransmitters
- **c** the reflex arc

HSW **4** Identify questions that science cannot address, and explain why these questions cannot be answered

Can you suggest something that your pancreas does?

Hormones

How do hormones control processes in your body?

When you are scared, impulses are sent to your adrenal glands and these produce adrenaline. This is a hormone that has many effects on your body. It makes blood vessels and the tubes leading to your lungs wider and also makes your heart beat faster. These effects prepare your body to fight or run away from something.

A *The effects of adrenaline make it a useful medicine. In some allergic reactions, the breathing tubes to the lungs start to g narrower and so adrenaline is used to keep them open.*

1 What is an endocrine gland?

2 Explain why hormones are often called 'chemical messengers'.

3 Transfer the information in Figure B to a summary table showing some human hormones, their endocrine glands, target organs and effects.

Hormones are produced and released by glands called **endocrine glands**. Hormones travel in the blood and act as 'chemical messengers', causing certain parts of the body to respond to their presence. For example, when there is a rise in human growth hormone during puberty, muscles and bones start to grow at a faster rate.

An organ that responds to a certain hormone is a **target organ**. Figure B shows some examples.

ResultsPlus
Watch Out!

Hormones go into the bloodstream and travel round the body where target organs detect them. A good way to remember when glucagon is secreted is that it sounds like 'glucose – gone'!

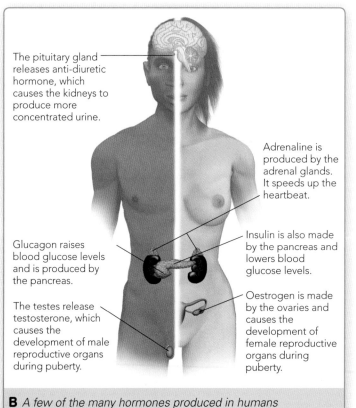

The pituitary gland releases anti-diuretic hormone, which causes the kidneys to produce more concentrated urine.

Adrenaline is produced by the adrenal glands. It speeds up the heartbeat.

Glucagon raises blood glucose levels and is produced by the pancreas.

Insulin is also made by the pancreas and lowers blood glucose levels.

The testes release testosterone, which causes the development of male reproductive organs during puberty.

Oestrogen is made by the ovaries and causes the development of female reproductive organs during puberty.

B *A few of the many hormones produced in humans*

Controlling blood glucose levels

Carbohydrates in your food are mainly digested into a sugar called **glucose**. After a meal the **concentration** of glucose in your blood goes up. When it gets above a certain concentration, your **pancreas** releases a hormone called **insulin**. The insulin is carried around your body in your blood.

Insulin affects certain cells, including those in your liver, which then take glucose out of the blood and convert it into **glycogen**. Glycogen acts as a store of glucose. This process means that the levels of glucose in the blood decreases.

When your blood sugar falls below a certain level, your pancreas releases another hormone, called **glucagon**. This is also carried in the blood and causes the cells in the liver to turn glycogen back into glucose, which is then released into the blood. The levels of glucose in the blood increase. Figure D shows how someone's glucose levels might change during the course of a day.

H

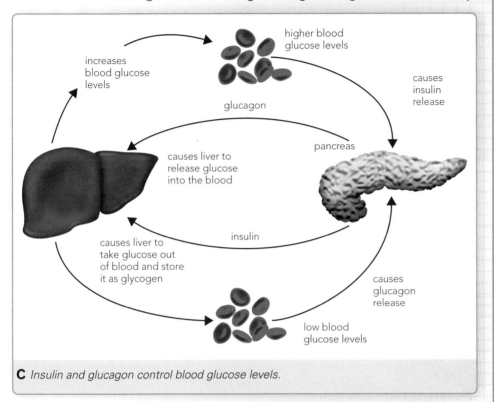

increases blood glucose levels

higher blood glucose levels

glucagon

causes insulin release

causes liver to release glucose into the blood

pancreas

causes liver to take glucose out of blood and store it as glycogen

insulin

causes glucagon release

low blood glucose levels

C *Insulin and glucagon control blood glucose levels.*

Skills spotlight

Figure D below shows the blood glucose concentrations of a person during the course of a day. With what sort of graph or chart should this data be presented? Explain your reasons.

Blood glucose concentration (mg/100 cm³)	Time of day
102	0700
118	0800
91	0900
87	1000
85	1100
83	1200
119	1300
114	1400
90	1500
87	1600
87	1700
83	1800
117	1900
103	2000
90	2100
78	2200
80	2300

D *Blood glucose levels throughout the day*

4 What stimulus is the pancreas responding to when it releases insulin?

5 What is the function of glycogen?

H 6 Figure C shows an example of homeostasis. What does this word mean?

7 Neurones and hormones both carry 'messages' around the body. How fast do you think messages are carried by hormones, compared to nerve cells? Explain your reasoning.

8 Explain the stages involved in the control of rising blood glucose levels by insulin.

Learning Outcomes

2.5 Recall that hormones are produced in endocrine glands and are transported by the blood to their target organs

2.6 Explain how blood glucose levels are regulated by insulin and excess blood glucose is converted to glycogen in the liver

H *2.7* Explain how blood glucose levels are regulated by glucagon causing the conversion of glycogen to glucose

HSW *11* Present information using scientific conventions and symbols

How is diabetes controlled?

The bees in Figure A have been trained to stick out their feeding tubes in response to certain gases, such as those given off by explosives. The bees could be used in an airport to warn of danger. It's not the first time that insect responses have been used in this way. About 2500 years ago an Indian doctor, Susruta, tested for diabetes by seeing if ants were attracted to people's urine.

1 What is diabetes?

People who have a disease called **diabetes** cannot control their blood glucose levels very well. Low blood glucose concentrations may cause unconsciousness. High blood glucose concentrations cause tiredness and can damage organs, such as the eyes. If blood glucose concentrations become too high, the kidneys get rid of it by putting it into the urine.

feeding tube

gases containing the gas the bees have been trained to detect

A *These bees are part of a vapour detection system.*

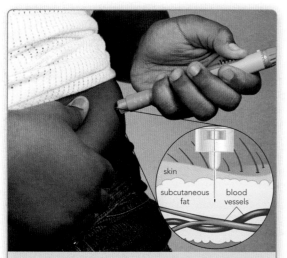

B *Insulin is often injected using a 'pen' three times a day during meals.*

Type 1 diabetes

Between 5 and 10% of diabetics have **Type 1 diabetes**, in which the pancreas does not produce insulin. This means that when blood glucose concentrations rise, the body cannot bring them back down to the correct levels. So, people with Type 1 diabetes usually inject insulin every day.

Insulin needs to be injected into the fat layer beneath the skin (the **subcutaneous fat** layer) because fat easily absorbs insulin. The insulin then spreads into blood vessels and is carried around the body in the blood.

To successfully control their disease, diabetics need to balance the factors that lower blood glucose levels with those that increase the levels. You can see from Figure C that the more exercise diabetics take, the less insulin they need. More insulin might be needed on some days than others.

2 Why must people with Type 1 diabetes inject themselves with insulin?

3 What may happen to someone who has Type 1 diabetes who forgets to inject insulin?

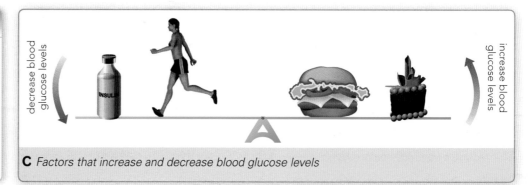

C *Factors that increase and decrease blood glucose levels*

Type 2 diabetes

Type 2 diabetes usually occurs when the cells in a person's body respond less well to insulin – the cells start to become 'resistant' to insulin, even though the hormone is being produced.

Scientists have discovered a number of risk factors that may help to cause Type 2 diabetes, including high-fat diets, lack of exercise, getting older and being obese (very overweight).

Doctors class people as obese if they have a **body mass index** (**BMI**) of over 30. A BMI gives an estimate of how healthy a person's mass is for their height. An obese person is likely to suffer health problems because of their mass.

$$BMI = \frac{\text{weight in kilograms}}{(\text{height in metres})^2} = \frac{kg}{m^2}$$

Type 2 diabetes can often be controlled by changing a person's diet and by increasing the amount of exercise they take.

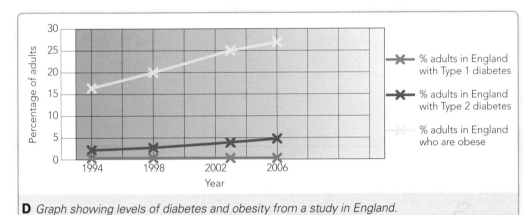

D *Graph showing levels of diabetes and obesity from a study in England.*

6 Evaluate the evidence in Figure D for correlation between obesity and Type 2 diabetes, but not Type 1 diabetes.

7 Suggest one way in which a person's diet will need to change if they have just been diagnosed with Type 2 diabetes.

8 Explain why the treatment of Type 2 diabetes may not require insulin but the treatment of Type 1 diabetes will.

4 Why does an increase in the amount of food that a diabetic eats also increase the amount of insulin they need?

5 Explain why taking exercise is helpful for controlling diabetes. (*Hint*: Think about respiration.)

Skills spotlight

In 1889, German scientist Oscar Minkowski (1858–1931) was studying digestion. He removed the pancreases from some dogs. The dogs became very ill and one of Minkowski's assistants noticed that ants were attracted to their urine. Explain how this gave Minkowski the idea that the pancreas was linked to diabetes.

ResultsPlus
Watch Out!

Type 1 diabetes occurs when insulin is not produced by the pancreas. Type 2 diabetes occurs when the body does not respond to the insulin produced.

Learning Outcomes

2.8 Recall that Type 1 diabetes is caused by a lack of insulin

2.9 Explain how Type 1 diabetes can be controlled, including the roles of diet and injection of insulin usually into the subcutaneous fat

2.10 Explain how, in Type 1 diabetes, the level of physical activity and diet affect the amount of insulin required

2.11 Recall that Type 2 diabetes is caused by a person becoming resistant to insulin

2.12 Explain how Type 2 diabetes can be controlled by diet and physical activity

2.13 Evaluate the correlation between obesity (including calculations of BMI) and Type 2 diabetes

HSW *2* Describe the importance of creative thought in the development of hypotheses and theories

Why do plant roots always grow downwards?

B1.17 Tropic responses

▶ Do seedling roots show a response to light?

Plant roots usually grow downwards but those spikes in Figure A are actually roots growing upwards, out of the ground! These are the roots of black mangrove plants, which live in waterlogged soil at the edges of the sea. The special upward growing roots get oxygen from the air, to allow the roots to respire.

A *Black mangroves have some roots that grow upwards.*

Like humans, plants respond to stimuli. Roots respond to the direction of gravity by growing towards it. You may have noticed another response of plants too; their stems and shoots grow towards the light.

B *Plant shoots growing towards the light.*

Charles Darwin (1809–1882) was not only interested in evolution. In 1880, he published a book called *The Power of Movement in Plants*. In it, he described many observations, including how barley seedlings responded to light coming from one direction. He discovered that the barley seedlings bent as they grew towards the light.

Your task

You are going to plan an investigation that will allow you to find out if plant shoots grow towards or away from different colours of light, or if they are unaffected. You are going to test the hypothesis that the growth of plant seedling shoots can be affected by the colour of light. Your teacher will provide you with some materials to help you organise this task.

Learning Outcomes

2.16 Investigate tropic responses

Build Better Answers

When completing an investigation like this, one of the skills you will be assessed on is your ability to *draw a conclusion*. There are 6 marks available for this skill. Here are two student extracts that focus on this skill. Other skills that you need for the assessment are dealt with in other lessons.

Student extract 1 — A basic response for this skill

It would be better to explain exactly what the data shows you – be specific.

> My conclusion is that light can make shoots bend. I can see this from my results. The table I found on the internet about flowering times and colours of light says that coloured light can make plants flower at different time.

This is a clear conclusion but it does not refer to the hypothesis.

This secondary evidence is irrelevant to this investigation.

Student extract 2 — A good response for this skill

You then need to give more detail of your scientific ideas to explain what has happened in your investigation.

> My conclusion is that shoots bend different amounts in different colours. In my practical the colours which had the greatest effect are blue and white. You can see this from my table of results where x% of the seeds grew towards blue light and y% grew towards white light. This proves my hypothesis that the growth of plant seedling shoots can be affected by the colour of light. This also matches my scientific knowledge that light effects the growth of seedlings. The blue light may cause the greatest response because it is the right wavelength for photosynthesis. White would cause a response because it contained blue light.

This answer also links the evidence to the hypothesis.

In order to access the top marks you need to refer to scientific ideas and whether the conclusion matches the scientific ideas or not.

 ResultsPlus

To access 4 marks
- Provide a conclusion based on all your collected evidence
- Refer to the original hypothesis in your conclusion
- Explain your conclusion using the evidence
- Describe any relevant mathematical relationships in your conclusion

To access 6 marks
You also need to refer to other relevant scientific ideas in your conclusion

How do plants use hormones to respond to stimuli?

The plant in Figure A was knocked over. The plant has grown a bend in the stem so that the leaves are back in the light. The Greek thinker Theophrastus (c371 BCE – c287 BCE) thought this bending was caused by sunlight removing fluid from the lit side of the plant, making it easier for the plant to grow on that side. This idea continued until the 17th century when scientists started to think plants could respond to stimuli.

A In the Middle Ages, plants that had bends in them to grow towards the light were given as treatment for snake bites because they looked like snakes.

Phototropism

Responding to a stimulus by *growing* towards or away from it is called a **tropism**. A tropism caused by light is a **phototropism**. A tropism *away from* a stimulus is a **negative tropism**. Plant roots are negatively phototropic.

Plant shoots are positively phototropic, to get enough light for **photosynthesis** (to make their own food).

1 Woodlice move away from light. Why is this not a tropism?

Hormones and phototropism

Like humans, plants produce hormones (also called **plant growth substances**) to respond to stimuli. Positive phototropism is caused by plant hormones called **auxins**.

Auxins are produced in the tips of a shoot, where they cause elongation of the cells. If a shoot is grown with light coming from only one direction, auxins move to the shaded side of the shoot. This makes the cells on the shaded side elongate more, which in turn causes the shoot to grow towards the light.

2 a What effect do auxis have on the cells in shoot tips?
b Why is this useful for the plant?

3 Give an example of a part of a plant that is negatively phototropic.

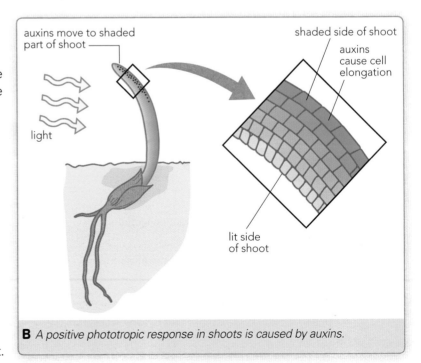

auxins move to shaded part of shoot

light

shaded side of shoot

auxins cause cell elongation

lit side of shoot

B A positive phototropic response in shoots is caused by auxins.

Auxins and gravitropism

Auxins are also found in root tips, where they have the *opposite* effect to that in shoots. In roots, auxins cause cells to stop elongating and this causes **positive gravitropism** or **geotropism** – growth towards the direction of gravity. It helps plant roots to anchor the plant in place and to reach moisture underground.

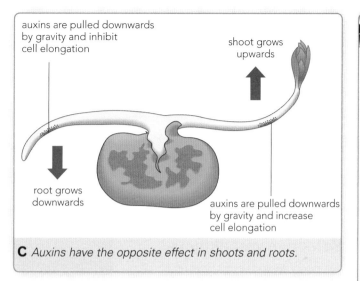

auxins are pulled downwards by gravity and inhibit cell elongation

shoot grows upwards

root grows downwards

auxins are pulled downwards by gravity and increase cell elongation

C *Auxins have the opposite effect in shoots and roots.*

Gibberellins

Auxins are not the only plant hormone. When a seed **germinates**, roots and a shoot (a new stem) start to grow. Some seeds need periods of darkness or cold before they will germinate. Once this period is completed, the seed releases plant hormones, called **gibberellins**. They cause the starch stored in a seed to be turned into sugars that the seed uses for energy to grow. Gibberellins also stimulate flower and fruit production in some plant species.

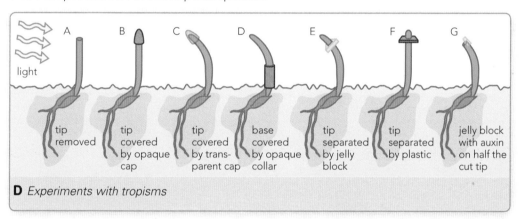

light

A	B	C	D	E	F	G
tip removed	tip covered by opaque cap	tip covered by transparent cap	base covered by opaque collar	tip separated by jelly block	tip separated by plastic	jelly block with auxin on half the cut tip

D *Experiments with tropisms*

5 State a stimulus that causes the production of gibberellins in a seed. ?

6 Two seeds of the same type are taken. Seed A is soaked in a solution of gibberellins. Seed B is soaked in water. Explain why seed A germinates first.

4 Explain why the shoots in the experiments shown in Figure D have or have not bent towards the light source. ?

Results Plus
Watch Out!

Auxins *retard* elongation in *roots* but *speed up* elongation in shoots.

Learning Outcomes

2.14 Explain how plant growth substances (hormones) bring about:
   a positivism phototropism in shoots
   b positive gravitropism (geotropism) in roots

2.15 Explain how auxins bring about shoot curvature using cell elongation

2.17 Analyse, interpret and evaluate data from plant hormone experiments, including the action of auxins and gibberellins

HSW **5** Plan to test a scientific idea, answer a scientific question or solve a scientific problem

How do humans make use of plant hormones?

Agent Orange is a weedkiller that was used in the Vietnam War (1959–1975). It destroyed the jungle so that the Americans could see enemy movements. It contains artificial auxins that cause plants to grow out of control and die.

A *Agent Orange is a powerful weedkiller.*

Selective weedkillers

Artificial auxin is still used as a **selective weedkiller** because it only makes plants with broad leaves (like daisies and dandelions) grow out of control and die. Plants with narrow leaves (like wheat and grass) are unaffected. Farmers can kill all the weeds in a field of a cereal crop (like wheat) without affecting their crop.

> **1** In what way are artificial auxin weedkillers 'selective'?
>
> **2** Why would a gardener use a selective weedkiller on a lawn?

B *A selective weedkiller kills the weeds but not the wheat. The weedkiller sprayer has missed a strip in this field.*

Rooting powder

Synthetic auxins are also used in **rooting powders**. Plant **cuttings** (parts of plants) are dipped in rooting powder, which makes them develop roots quickly. Large numbers of the same plant can be produced quickly using cuttings, compared to growing plants from seed.

> **3** Many plant cuttings develop roots without using rooting powder. What is the advantage of using rooting powder on cuttings?

C *The cuttings on the right were dipped in rooting powder.*

Seedless fruit

Some seedless fruits are produced using plant hormones. The flowers are sprayed with plant hormones that cause fruits to develop but not their seeds.

Other plants, like some varieties of grape, have naturally seedless but small fruits. The fruits are sprayed with gibberellins to increase their size.

D Most seedless grapes have been sprayed with plant hormones to make the fruits bigger.

Fruit ripening

Plant hormones naturally control the ripening of fruits. Farmers can make use of this and control when and how ripening occurs. For example, plant hormones are sprayed onto:

- fruit trees to stop the fruit falling off. This stops fruits falling and becoming damaged and also allows the fruit to grow bigger.
- fruit trees to speed up ripening. All the fruit ripens together and can be picked in one go.
- unripe fruit to make them ripe. The fruit reaches shops in a 'just ripened' condition.

E Unripe, green bananas from the Caribbean are ripened in the UK using a natural plant hormone – a gas called ethene or ethylene.

Results Plus
Watch Out!

When considering the uses of plant hormones, realise that 'describe how thay are used' is a different question to 'explain why they are used'.

Skills spotlight

Decisions to use science in food production are often based on economics – whether the use of science can increase the profit. Many UK apple growers pick apples that are unripe and ripen them later with plant hormones. Suggest why they do this. How would a farmer judge whether this is a good decision?

4 How do you think weeds on a playing field are killed without destroying the grass? Explain your answer.

5 There is a huge range of exotic fruit in supermarkets. Twenty-five years ago the range was much smaller. Explain why you think this is so.

6 In autumn, farm shops sell apples that have been recently picked. Suggest three ways in which plant hormones may have been used in the production of these apples and explain why farmers may use these methods.

Learning Outcomes

2.17 Analyse, interpret and evaluate data from plant hormone experiments, including the action of auxins and gibberellins

H 2.18 Demonstrate an understanding of the uses of plant hormones, including:
   a selective weedkillers
   b rooting powder
   c seedless fruit
   d fruit ripening

HSW 13 Describe the social, economic and environmental effects of decisions about the uses of science and technology

>>>>>>>>>>> What famous soft drink has been banned various times in France, Norway and Denmark?

⠿ **How can drugs affect us?**

Many plants produce chemical substances that taste unpleasant or are poisonous, to stop herbivores eating them. Some of these chemical substances affect people too. Many medical drugs are based on these chemical substances, such as morphine which comes from poppies. Many recreational drugs, including heroin, cocaine and caffeine, also come from plants.

A *Coffee plants evolved to produce caffeine in their berries which kills insects that try to eat them.*

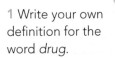

1 Write your own definition for the word *drug*.

2 Some people say computer games are a drug. Is this a correct use of the word? Explain your answer.

What is a drug?
Our bodies make many chemical substances. Some of these (e.g. hormones, neurotransmitters) coordinate how we behave and respond to stimuli. Adding other chemical substances to our bodies can interfere with the chemicals that our bodies make and change the way they work. Any chemical substance that changes the way in which the body works, including our behaviour, is called a **drug**. Some drugs particularly affect the central nervous system and change our psychological behaviour in the way we feel, think and act.

Different types of drugs
We group drugs by the effects they have on us.

ResultsPlus
Watch Out!

When a reaction time is less, the reaction is faster. When a reaction time is more, the reaction is slower.

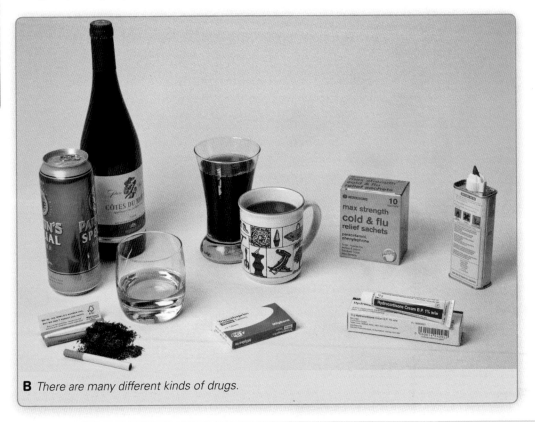

B *There are many different kinds of drugs.*

- A **narcotic** is a drug that makes us feel sleepy.
- We feel pain when electrical impulses from a swollen or damaged area of the body are sent via neurones to the brain. **Painkillers**, such as morphine, block some of these nerve impulses, so we feel less pain.
- **Hallucinogens** change the way the brain works, particularly how we respond to what we see, hear and feel. Drugs like LSD can distort our senses of colour, time and space.
- **Stimulants** like caffeine increase the speed of **neurotransmission** (of nerve impulses) across synapses. This speeds up **reaction times** (the time it takes for the body to respond to an outside stimulus, such as a sound).
- **Depressants**, such as alcohol, slow down the activity of neurones in the brain and can help us relax.

Any drug that is used to make people feel a certain way is a recreational drug. Alcohol is a recreational drug. Some drugs are medicines – they help to limit damage caused by diseases or injuries.

Some drugs are **legal** (e.g. caffeine in cola drinks) but some are only legal to buy at certain ages (e.g. alcohol at 18). Other drugs are **illegal** to buy because they can have dangerous effects. Some illegal drugs (e.g. heroin) are used in carefully controlled conditions in hospitals.

Most drugs can be **addictive** which means that people become dependent on the drug and feel that they cannot function properly without it.

> 3 Look at Figure B.
> a Sort the drugs into medical drugs and recreational drugs.
> b Explain your choices.
>
> 4 Write down all the categories of drugs mentioned on these pages and give examples of each. Indicate which are legal drugs and which are illegal.
>
> 5 Not all people react to drugs, such as caffeine, in the same way. Should governments be allowed to ban drinks that they think can cause harm? Suggest reasons for your answer.
>
> 6 Alcohol increases reaction times. Explain fully how it does this.

C Drinks containing high doses of caffeine have been banned in some countries because they have been associated with several deaths.

Learning Outcomes

3.1 Define a drug as a chemical substance, such as a narcotic or hallucinogen, that affects the central nervous system, causing changes in psychological behaviour and possible addiction

3.2 Describe the general effects of:
- a painkillers that block nerve impulses, including morphine
- b hallucinogens that distort sense perception, including LSD
- c stimulants that increase the speed of reactions and neurotransmission at the synapse, including caffeine
- d depressants that slow down the activity of the brain, including alcohol

HSW 13 Explain how and why decisions that raise ethical issues about uses of science and technology are made

Why can't athletes use many common medicines?

B1.21 Reaction times and drugs

How do drugs affect your reactions?

Alain Baxter was the first-ever British skier to win a medal at a Winter Olympics (held in 2002). A few days after returning home to Scotland he was told he had to hand his medal back because he'd failed a drugs test. He said that he'd taken a cold remedy not knowing that it contained a banned stimulant. An appeal committee later said that, although they believed his story, they would not give him his medal back.

The drug that Alain took was a type of amphetamine. These drugs have been used in cold remedies since the 1930s. They help to unblock your nose. However, they can have other effects including making people more alert and decreasing their **reaction times**. It is this effect that has put them on the list of banned substances for athletes.

A Alain Baxter

B Sprinters need to have fast reaction times.

Your task

You are going to plan an investigation that will allow you to find out about reaction times and how they are changed by caffeine. You are going to test the hypothesis that the caffeine in cola drinks decreases reaction times. Your teacher will provide you with some materials to help you organise this task.

Learning Outcomes

3.3 Investigate reaction times

When completing an investigation like this, one of the skills you will be assessed on is your ability to *evaluate your conclusion.* There are 4 marks available for this skill. Here are two student extracts that focus on this skill. Other skills that you need for the assessment are dealt with in other lessons.

Student extract 1 — A basic response for this skill

Explain what you would be looking for on the Internet or in a textbook – be specific.

This is a good explanation of how you could change the method to make sure that any results you gathered were good quality.

> I think that my conclusion is strong because my own results showed that drinking the caffeine did change the speed of the reaction time. I think that if I found some more information on the internet this would help to strengthen my conclusion. I could also take a reading before people drink anything so that we had a basic idea of what their reaction time is like, and we could not tell people which cola they were drinking so they did not think that one would increase their reaction times and try harder.

You need to evaluate the conclusion using all the evidence so make sure that you include any secondary evidence you found and whether this supported your conclusion.

Student extract 2 — A good response for this skill

This is good because it explains clearly one approach that you could take.

> I think that my conclusion was a strong one because the results from my investigation showed a clear change after the caffeine had been drunk and the table of results I found on the internet about the effect of coffee on reaction times also showed a clear change. This is what I would expect from the scientific knowledge. To extend the practical and find out more I would take more readings. I would also look on the internet to see if there is any data available on caffeine drinks and how these affect reaction times – possibly in the context of driving.

Say what your scientific knowledge is and what you would expect the results to show if your science was correct.

Think about other questions you could ask and how you could test them.

To access 2 marks
- Evaluate how well all your evidence supports your conclusion
- Suggest how your evidence can be improved to strengthen your conclusion

To access 4 marks
You also need to:
- Evaluate how well other scientific ideas support your conclusion
- Suggest how the investigation could be extended to support your conclusion

The damage caused by smoking

How can smoking cigarettes harm you?

Some people are so addicted to smoking that they cannot stop, even though the damage to blood vessels caused by smoking means they need to have their toes removed.

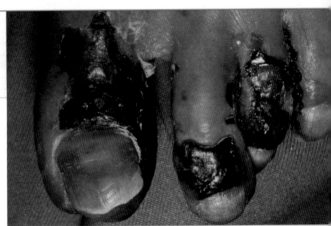

A *The black areas are dead tissue. The blood vessels to the cells are so damaged that the cells can't get the substance they need to stay alive.*

Damage caused by tar

Tobacco smoke contains many chemical substances that damage living tissue. The sticky **tar** in the smoke contains chemical substances called **carcinogens**. These can cause cancers, which develop most often in the lungs and mouth. An estimated 42 000 people in the UK die each year from cancers related to smoking.

> **1 a** Describe the differences between the lungs in Figure B.
> **b** Explain why they are different.

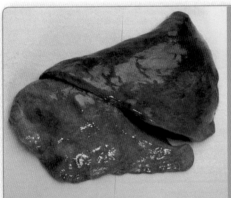

B *The human lung on the left is from a non-smoker; the one on the right is from a smoker. The white parts are cancer.*

Damage caused by carbon monoxide

Carbon monoxide is a gas that is poisonous because it reduces the amount of oxygen that red blood cells can carry. It is found in tobacco smoke. Cells all around the body need oxygen to work properly. A lack of oxygen to active muscles can cause pain, such as in the legs when walking.

Carbon monoxide also makes blood vessels narrower. Therefore, body cells supplied by those blood vessels get even less oxygen and may die. Dead tissues easily get infected and must then be removed. If the heart muscles get too little oxygen, this causes a heart attack which may kill. Each year in the UK, smoking causes about 30 000 deaths from circulatory diseases (diseases of the heart and blood vessels).

Carbon monoxide and other gases in tobacco smoke can also damage lung tissue. This causes respiratory diseases such as emphysema and chronic bronchitis. Each year in the UK, smoking causes another 30 000 deaths from lung diseases.

> **2** List the effects of carbon monoxide on the body.
>
> **3** Tobacco smoke only goes into the lungs. Explain why smoking can cause diseases in all parts of the body.

The problem with nicotine

A drug in tobacco smoke is **nicotine**. It is the addictive part of tobacco smoke and makes it difficult to give up smoking. Addiction may start after smoking just four cigarettes but may last for years, even after giving up.

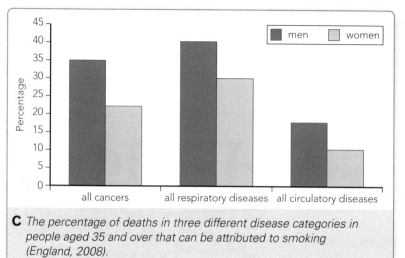

C *The percentage of deaths in three different disease categories in people aged 35 and over that can be attributed to smoking (England, 2008).*

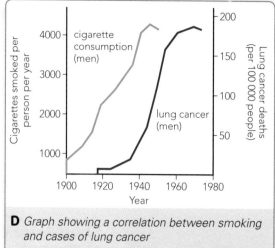

D *Graph showing a correlation between smoking and cases of lung cancer*

Evaluating evidence

It can take many years for smoking to create sufficient damage in the body to cause death. During that time a person will be exposed to many other substances. This makes proving that smoking was the cause of death (and not exposure to other substances) more difficult.

There have been many studies on the link between smoking and disease. We have to evaluate these studies to see if their conclusions are reliable. Long-term studies that look at individual people for many years and studies that include large numbers of people are more likely to give us reliable evidence because they help to average out the effects of normal variation in results.

4 Suggest why, even when people know smoking is harmful, they can't give it up.

5 Use the data given on these two pages to calculate the total number of deaths in the UK each year from cancers, respiratory diseases and circulatory diseases caused by smoking.

6 Look at Figure C.
a Describe what the graph shows.
b Suggest a reason for the difference between men and women.
c Not everyone agrees that smoking is the cause of these proportions of deaths. Explain what you would do to evaluate the evidence in this graph.

7 To help show the link between smoking and lung cancer, scientists collected the data in Figure D. Describe what the graph shows and suggest why there is a delay in the effect of smoking on lung cancer deaths.

Maths skills

Figure C is a type of **bar chart**. Bar charts represent qualitative or discontinuous data. The size of the bars represents the size of each category.

ResultsPlus
Watch Out!

Nicotine is the chemical that causes addiction. Tar, carbon monoxide and other substances cause the physical damage.

Skills spotlight

There is evidence that tobacco companies have known for many years that smoking damages health, but they kept those reports secret. What reasons might the company have for this decision?

Learning Outcomes

3.4 Explain the effects of some chemicals in cigarette smoke, including:
   a nicotine as an addictive drug
   b tar as a carcinogen
   c carbon monoxide reducing the oxygen-carrying ability of the blood

3.5 Evaluate data relating to the correlation between smoking and its negative effects on health

HSW 13 Explain how and why decisions that raise ethical issues about uses of science and technology are made

>>>>>>>>>>>>>>>>>>>>>>>> How many people die each year due to alcohol?

The effects of alcohol

What does alcohol do to us?

Mark Shields died after binge drinking at a pub with his friends to celebrate his 18th birthday. In 40 minutes he had drunk three lagers, three double liqueurs and five double whiskies. He died of acute alcohol poisoning.

1 Explain why alcohol is such a popular drug.

2 Why is it important not to drive after drinking alcohol?

3 List three short-term effects of drinking alcohol.

Short-term effects of alcohol

Alcohol is quickly absorbed by cells and so affects all the organs of the body. It first slows down activity in the brain and nervous system. This affects reaction times so the person takes longer to respond. People drink alcohol because it reduces negative feelings and makes them feel happier. It can also lower **inhibitions**, so users are more likely to do things that they wouldn't normally do, including taking greater risks.

A

THE Herald

BIRTHDAY BOY DIES AFTER 40 MINUTES OF BINGE DRINKING

A boy died yesterday after a 40-minute drinking binge in Hexham.

Kimberly Hoppe

Teenager Mark Shields' blood alcohol level was found to be over 6 times the legal alcohol limit after he spent only 30–40 minutes at a bar drinking. In that time Shields consumed 3 pints of beer, 3 double Aftershock shots and 5 double whiskies.

Shields' friends thought he d_ seem that drunk and took _ home to bed. The next mornin_ 18th birthday, his father foun_ in his bed dead.

Skills spotlight

An argument is a discussion in which evidence is presented for or against an idea and then used to reach a decision for or against. Use scientific evidence to construct an argument for or against raising the legal age for buying alcohol to 21.

The more alcohol there is in the body the slower the nervous system becomes. This causes blurred vision and affects coordination, making it more difficult to do simple physical or mental tasks. Very large amounts can cause unconsciousness and possible death by choking on vomit. It can also slow the nervous system down so much that breathing stops.

4 Describe two long-term effects of drinking alcohol.

5 Explain why the liver is most likely to be affected by heavy drinking.

B Even though alcohol is legal, it can have harmful effects on the body.

Long-term effects of alcohol

Drinking alcohol frequently can damage all the organs of the body. Alcohol in the blood is taken to the liver to be broken down. Long-term heavy drinking can cause **cirrhosis** of the liver, where normal tissue is destroyed so the liver cannot function properly and this can lead to death. Heavy drinking over a long time will also damage the brain, affecting learning and memory or causing a blood clot in the brain.

Alcohol can also be addictive; people who become dependent on its effects are called alcoholics.

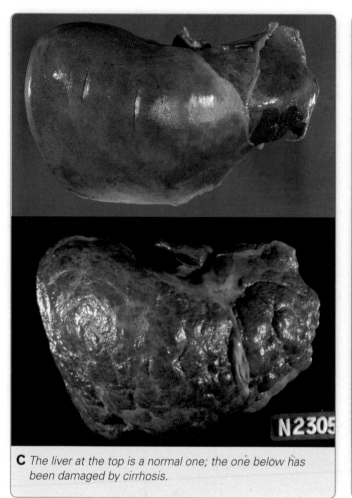

C *The liver at the top is a normal one; the one below has been damaged by cirrhosis.*

D *Many people die as a result of the effects of alcohol on a driver's reaction time.*

The cost to society of alcohol

As alcohol affects the way we behave, it not only affects us but also the people around us. Violence towards others is more common when people are drunk. Accidents, such as falling, are also more likely when someone is drunk, as is suicide.

Each year in the UK, there are over 9000 deaths that can be directly related to alcohol. However, it has been estimated that up to 40000 deaths a year may be the result of alcohol and that up to 5% of the money spent on health in the UK each year is used to treat people who have been drinking excessively.

6 Explain why there may be as many as 40000 alcohol-related deaths per year in the UK but only about 9000 are recorded as officially being due to alcohol.

Results Plus
Watch Out!

If you are describing the effects of alcohol, be clear about the short-term and long-term effects on the individual or on society.

Learning Outcomes

3.6 Evaluate evidence of some harmful effects of alcohol abuse:
   a in the short term – blurred vision, lowering of inhibitions, slowing of reactions
   b in the long term – liver cirrhosis, brain damage

HSW *11* Present information, develop an argument and draw a conclusion using scientific, technical and mathematical language

>>>>>>>>>>>>>>>>>>>>>>>>

Should alcohol abusers get liver transplants?

Should everyone get the same medical treatment?

Liam died aged 23 from liver damage after 10 years of binge drinking. He was refused a liver transplant and died in hospital because he could not prove that he could give up alcohol. The chart on the right shows the results of a survey where people were asked what they thought of gving liver transplants to alcoholics. Which way would you vote?

Transplanting organs

In a **transplant**, a healthy organ such as a heart or liver is taken from one body (the **donor**) and put into a patient to replace an organ that no longer functions properly.

A *Yes, alcoholics should be given liver transplants: 53%. No, alcoholics should not be given liver transplants: 47%.*

Difficult decisions

Each year more than 1000 people in the UK die waiting for transplants because there are not enough donor organs. Doctors must decide which patients to operate on.

Doctors use scientific **criteria** to make decisions based on the likelihood of success. These include whether the patient and donor:

- have similar tissues – the surfaces of cells come in different 'types' and the closer the match the more likely the transplant will be successful
- are similar ages – a child's organ is less likely to be successful in an adult and vice versa
- are geographically close – the quicker the organ is transplanted the more likely the operation will be successful
- how ill the patient is – a very ill patient is less likely to survive an operation than a more healthy one.

> **1** Why must doctors choose who gets a transplant?
>
> **2** Why must all doctors use the same criteria for making decisions?

Other criteria are based on what people think is right or wrong. These are ethical criteria and they are often based on patients' lifestyles. People who abuse alcohol may get a disease called cirrhosis, which destroys their liver. Some hospitals will not give liver transplants to patients who abuse alcohol unless they can stay off alcohol for more than six months outside hospital.

> **3** How might the ethical criteria for liver transplants differ from hospital to hospital?

Being **clinically obese** (very overweight) can damage the heart. So some hospitals will not perform heart transplants on obese people unless they can stick to a weight loss diet.

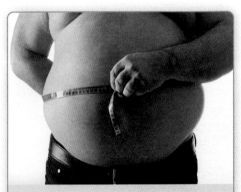

B *Being very obese (see B1.16) can increase the risks of problems during major operations such as heart surgery.*

An **ethical decision** uses ethical criteria (standards) to reach an answer that most people think is right or fair. Ethical decisions are hard because different people have different standards by which they judge what is right and what is wrong. For example, some people think that certain lifestyles that cause diseases (e.g. alcohol addiction) are not the person's fault, so they should have the same chances for treatment as everyone else. Other people think that lifestyle is a choice – you can choose whether or not to drink or smoke heavily and risk disease, so you shouldn't be allowed the same chance for treatment.

Other ethical issues

Some donors would like to choose who should or should not get their organs. Some people think that it should be possible to sell organs from relatives who have recently died or from themselves (e.g. you have two kidneys but can live with only one).

Results Plus
Watch Out!

When writing about ethical issues, try to consider both sides of the discussion. Avoid generalisations.

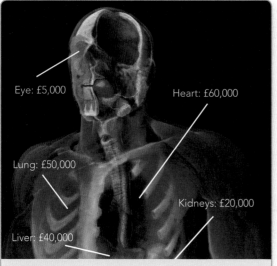

Eye: £5,000
Heart: £60,000
Lung: £50,000
Kidneys: £20,000
Liver: £40,000

C *Organs are often sold illegally for illegal transplants.*

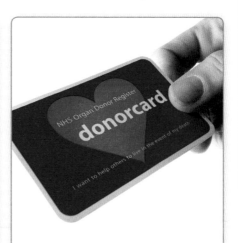

D *In the UK, carrying a donor card shows that you give permission for your organs to be used after your death to save the lives of others.*

Skills spotlight

In the UK, people must give permission for their organs to be used after death but other countries have an 'opt-out' policy for organ donation. Suggest possible reasons for these two different decisions.

4 Explain the difference between scientific and ethical decisions.

5 Joe Snow (a white man) doesn't want his organs to go to a black person. Is it ethical to grant Joe Snow's wishes? Explain your answer.

6 Liam was so ill with liver disease that he died before he could prove he could stay off alcohol for six months. Do you think the decision of the hospital not to give him a transplant was ethical? Explain your answer.

7 Suggest, with reasons, how you think the health service should decide who gets organ transplants.

Learning Outcomes

3.7 Discuss the ethics of organ transplants, including:
   a liver transplants for alcoholics
   b heart transplants for the clinically obese
   c the supply of organs

HSW 13 Explain how and why decisions that raise ethical issues about uses of science and technology are made

How do we get infections?

In the UK, transplant organs and blood are screened (tested) for certain viruses and bacteria to make sure a transplant patient won't get an infection from them. The patient is also treated with large doses of antibiotics to prevent bacterial infections during the operation.

A Dame Anita Roddick, founder of the Body Shop company, died after having cirrhosis of the liver caused by a virus called hepatitis C. She probably got it during a blood transfusion before screening for the virus was introduced in 1991.

Pathogens and disease

Microorganisms are organisms that are too small to see without magnification. There are millions of species of microorganisms, some of which live on or in our bodies. A few are **pathogens** because they cause **infectious diseases** when they are passed from an infected person to someone who is not infected.

Many pathogens are **bacteria**, such as those that cause cholera, food poisoning, dysentery and tuberculosis (TB). Other diseases are caused by **viruses** (e.g. influenza, mumps, measles and AIDS). A few pathogens are **fungi**, such as the athlete's foot fungus, and some are **protoctists**, such as the **protozoan** that causes malaria.

1 Write definitions for these terms in your own words:
a pathogen b infectious disease.

2 Describe how cirrhosis of the liver can be caused by:
a an infectious disease b a non-infectious disease.

3 Draw up and complete a table with the following headings to show the different kinds of pathogens. Leave the right-hand column blank for now.

Type	Examples	
bacteria		

Skills spotlight

Scientific developments have benefits, risks and drawbacks. One problem with transplants is that until there is a test for a particular infection, such as hepatitis C, we cannot be certain if we are transferring that infection along with tissue in a transplant. Describe the benefits, risks and drawbacks of giving a blood transfusion or transplant to a patient.

How pathogens pass between people

For pathogens that cause human diseases to spread they need to pass from an infected person to someone who is not infected. They do this in many different ways, sometimes through direct contact between people and sometimes carried by **vectors** (animals that pass pathogens from one person to another).

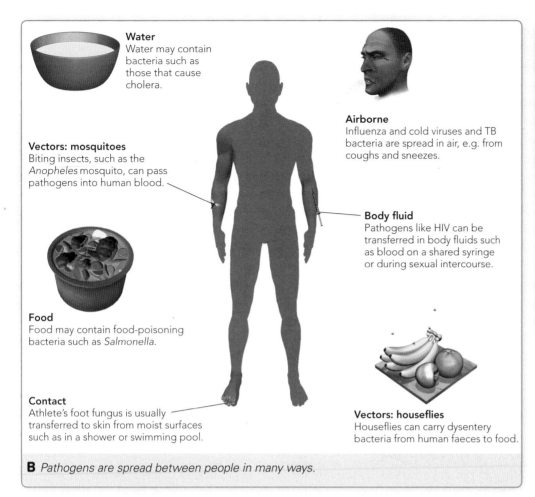

Water
Water may contain bacteria such as those that cause cholera.

Airborne
Influenza and cold viruses and TB bacteria are spread in air, e.g. from coughs and sneezes.

Vectors: mosquitoes
Biting insects, such as the *Anopheles* mosquito, can pass pathogens into human blood.

Body fluid
Pathogens like HIV can be transferred in body fluids such as blood on a shared syringe or during sexual intercourse.

Food
Food may contain food-poisoning bacteria such as *Salmonella*.

Contact
Athlete's foot fungus is usually transferred to skin from moist surfaces such as in a shower or swimming pool.

Vectors: houseflies
Houseflies can carry dysentery bacteria from human faeces to food.

B *Pathogens are spread between people in many ways.*

C *If a female Anopheles mosquito feeds on human blood that contains the malaria protozoan it can pass the microorganism to the next human that it feeds on.*

Results Plus
Watch Out!

Pathogens cause disease, but there are several types and they can be transmitted in many different ways.

4 Give two examples of vectors.

5 Use the information in Figure B to complete the third column of your table to show how your examples of pathogens are passed between people.

6 During the 2009 swine flu epidemic, people were told to 'catch it, bin it, kill it'. The advice was to stress the importance of using a tissue when you sneeze. Use your knowledge of how pathogens spread to explain the importance of this advice.

Learning Outcomes

3.8 Recall that infectious diseases are caused by pathogens

3.9 Describe how pathogens are spread, including:
  a in water, including cholera bacterium
  b by food, including *Salmonella* bacterium
  c airborne (eg sneezing), including influenza virus
  d by contact, including athlete's foot fungus
  e by body fluids, including HIV
  f by animal vectors, including:
    i housefly: dysentery bacterium
    ii *Anopheles* mosquito: malarial protozoan

HSW 12 Describe the benefits, drawbacks and risks of using new scientific and technological developments

How can we reduce the risk of being infected?

The Komodo dragon has a weak bite, so it bites its prey and then follows for a few days as the prey bleeds to death. It was thought that bacteria in the lizard's mouth helped to poison the prey and kill it more quickly, but recent research shows that the lizard's saliva also includes snake-like venom which speeds up the bleeding.

A *The Komodo dragon is a 3-metre long lizard that kills and ea large prey.*

---

1 Give two physical barriers of the human body to infection.

2 Explain why wiping surfaces with antiseptic where fresh meat has been prepared is helpful.

3 a Which of the following diseases could be treated with antibiotics: influenza, cholera, *Salmonella* food poisoning, HIV, colds?
b Explain your answer.

Defence against invasion

Animals, including humans, have many different ways to protect themselves against invasion by pathogens. Some of these are **physical barriers** that stop pathogens getting into the body. We also have **chemical defences** that help to kill pathogens before they can harm us.

Using antiseptics

We can reduce infection by preventing the pathogen that causes it from getting into the body. Chemical substances that kill microorganisms outside the body are called **antiseptics**.

Antiseptics can help prevent pathogens getting into an open wound. They are also important where there are a lot of pathogens, such as around the toilet, or where fresh meat, which may contain food-poisoning bacteria, is being prepared.

Plants also defend themselves against attack. They use chemical substances that are **antibacterials** to prevent pathogens from causing damage. We use some of these for their antiseptic properties. For example, witch hazel from *Hamamelis* is used in some aftershaves to help prevent infection through cuts in the skin and mint in toothpaste may do more than give a fresh taste.

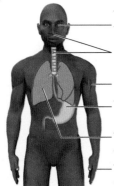

Tear glands make a liquid containing enzymes called **lysozymes** that kill microorganisms.

Hairs in the nose filter out dust that might carry pathogens. **Mucus** in the nose, throat and breathing passages traps microorganisms.

If the skin is broken, the blood clots to block entry by pathogens.

The stomach makes hydrochloric acid to kill harmful microorganisms in food.

The tube in the lungs also produces mucus. Tiny hairs called **cilia** sweep out the mucus and microorganisms trapped in it.

The skin forms a protective barrier. Sweat glands in skin make chemical substances that kill harmful microorganisms.

B *How the human body is protected from pathogens*

Using antibiotics

If pathogens get into the body, then we need a way of killing them without killing the patient. **Antibiotics** are chemical substances that kill or prevent the growth of bacteria and some fungi but not human cells. Antibiotics that only affect fungi, such as nystatin, are called **antifungals**. Those that only affect bacteria, such as penicillin, are called antibacterials. Unfortunately, antibiotics do not kill viruses.

Resistance

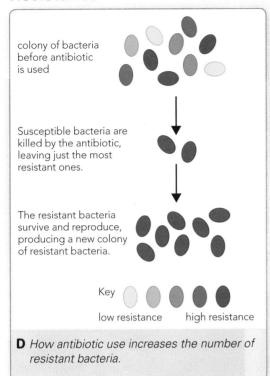

colony of bacteria before antibiotic is used

Susceptible bacteria are killed by the antibiotic, leaving just the most resistant ones.

The resistant bacteria survive and reproduce, producing a new colony of resistant bacteria.

Key

low resistance   high resistance

D *How antibiotic use increases the number of resistant bacteria.*

Individual bacteria in a population show **variation**, so some will naturally be more **resistant** to an antibiotic and take much longer to be killed than others. When the antibiotic is first used, the less resistant bacteria are killed first and the patient starts to feel better. However, the more resistant bacteria are still there and can cause infection again if the patient stops taking the antibiotics. And if these more resistant bacteria reproduce and spread to other people, they can cause an infection that cannot be treated with that antibiotic. So, using antibiotics encourages bacterial resistance to become more common. **MRSA** is a bacterium that is resistant to many antibiotics.

Skills spotlight

C *Fleming's culture plate of bacteria with the mould colony at the top.*

In 1928, no one knew about antibiotics. Alexander Fleming noticed a mould growing on one of his bacterial culture plates. No bacteria were growing around the mould. He suggested that the mould was releasing a substance that killed the bacteria – an antibiotic. How did Fleming's observations provide evidence for testing this hypothesis?

Overusing antibiotics also allows resistance to develop because each antibiotic kills many types of bacteria, not just those it is being used against. This means that resistant bacteria of all those different types will be left behind when the antibiotic treatment stops.

H 4 Why should you always finish a course of antibiotics, even if you are feeling better after taking them for a short time?

H 5 Explain in your own words how bacteria resistant to many antibiotics can develop.

6 Explain as fully as possible how we protect ourselves against diseases.

ResultsPlus
Watch Out!

Antiseptics, antibiotics and antibodies are similar sounding words and can be easily confused. Check the meaning of each.

Learning Outcomes

3.10 Explain how the human body can be effective against attack from pathogens, including:
   a physical barriers – skin, cilia, mucus *b* chemical defence – hydrochloric acid in the stomach, lyozymes in tears

3.11 Demonstrate an understanding that plants produce chemicals that have antibacterial effects in order to defend themselves, some of which are used by humans

3.12 Describe how antiseptics can be used to prevent the spread of infection

3.13 Explain the use of antibiotics to control infection, including:
   a antibacterials to treat bacterial infections
   b antifungals to treat fungal infections

H *3.14* Evaluate evidence that resistant strains of bacteria, including MRSA, can arise from the misuse of antibiotics

HSW *2* Describe how data is used by scientists to provide evidence that increases our scientific understanding

>>>>>>>>>>>>>>>>>>> Why did some dead bodies have bricks in their mouths in 16th century Venice?

B1.27 Antiseptics

How do antiseptics affect bacteria?

Before the 19th century, people had all sorts of ideas about what caused infectious diseases (those that are spread from person to person). These ideas included foul-smelling air, gods and even vampires.

Now that we know that microorganisms cause infectious diseases, we have found ways of killing the microorganisms to prevent their spread and to cure people.

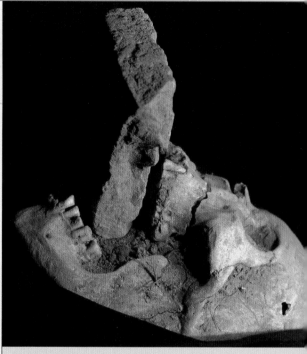

A *In 16th century Venice, bricks were pushed into the mouths of bodies that were suspected of being vampires. This was to ensure they starved and couldn't rise at night to attack people and cause diseases like the plague.*

B *Listerine® was originally designed to be used during surgery and was named after Joseph Lister. Later it was used as a mouthwash.*

Joseph Lister
(1827– 1912) was a Scottish surgeon who first showed that using antiseptics could stop infections. Antiseptics are now found in all sorts of products that we use every day.

Your task

You are going to plan an investigation that will allow you to find out how the concentration of an antiseptic affects its ability to kill bacteria. You are going to investigate the hypothesis that the higher the concentration of an antiseptic, the better it is at killing bacteria. Your teacher will provide you with some materials to help you organise this task.

Learning Outcomes

3.15 Investigate the effects of antiseptics or antibiotics on microbial cultures

Build Better Answers

When completing an investigation like this, one of the skills you will be assessed on is your ability to *assess the risks in an investigation*. There are 4 marks available for this skill. Here are two student extracts that focus on this skill. Other skills that you need for the assessment are dealt with in other lessons.

Student extract 1 | A basic response for this skill

The greatest risk is actually from any bacteria you might accidentally grow. These could be bacteria which come from dirty hands or coughs and sneezes.

> The greatest risk is the risk from the bacteria in the plate I used as part of this investigation. To control this risk I will make sure that I wash my hands after I have finished the investigation.

This is a good suggestion – but ideally you would have more than one.

Student extract 2 | A good response for this skill

Risks like this are not relevant to the practical.

> There are lots of risks in this practical. Here is a list of them: risk from bacteria in the plates or any that are accidently added to the plates and then grown in the incubator, risk from any spills on the floor, risk from the electricity sockets and gas taps in the laboratory. To help control these risks I can make sure that I wash my hands before and after the practical, clean the benches and equipment properly with alcohol, not open the plates once they are set up, incubate the plates at a low temperature, mop up spills, don't play with the gas taps or sockets.

Making a list of the hazards and then how you are going to control them is a good way to tackle this section of the investigation.

Because this is not a relevant risk you don't need to suggest how to manage it.

 ResultsPlus

To access 2 marks
- Identify a relevant risk which is specific to the investigation
- Suggest how to deal with that risk

To access 4 marks
- Identify most of the relevant risks which are specific to the investigation
- Describe how to manage all the risks involved

How are organisms dependent on each other for food?

Athlete's foot fungus feeds on the skin of the feet of humans. If there weren't any people, would athlete's foot fungus die out too?

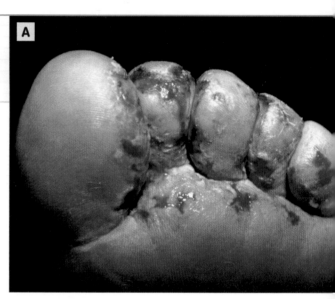

A

1 Write down two examples of:
a producers
b consumers.

2 Use Figure B to write down three different food chains.

3 Explain in terms of energy transfer why all food chains start with a producer.

Food webs

All organisms need food. Some organisms are **producers** and make their own food, such as green plants, which use photosynthesis. The rest get their food from other organisms – either as **primary consumers** (by eating plants) or **secondary consumers** (by eating primary consumers).

Food chains show what eats what. Organisms that feed at the same level in a food chain are in the same **trophic level**. Food chains from a habitat can be joined together into a **food web**, which shows the **feeding relationships** between the different organisms. You can see that organisms in an area depend on each other for food – they are **interdependent**. As the numbers of one organism change, other organisms are affected, so the relationships among the organisms are always changing. We call this a dynamic relationship. Dynamic means changing.

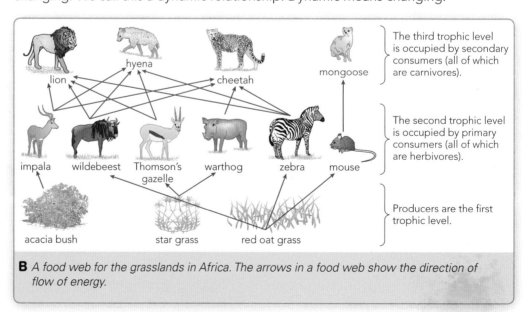

B A food web for the grasslands in Africa. The arrows in a food web show the direction of flow of energy.

Energy transfers in food chains

Inside an organism, energy stored in food is released during **respiration**. Some of this energy is transferred into **biomass** (substances that form tissues) as the organism grows. The energy stored in biomass will be transferred to the next organism in the food chain when it is eaten.

However, some of the energy that is released in respiration is transferred into forms of energy that are not useful. This energy is transferred to the environment and is therefore wasted and is not available to organisms at the next trophic level.

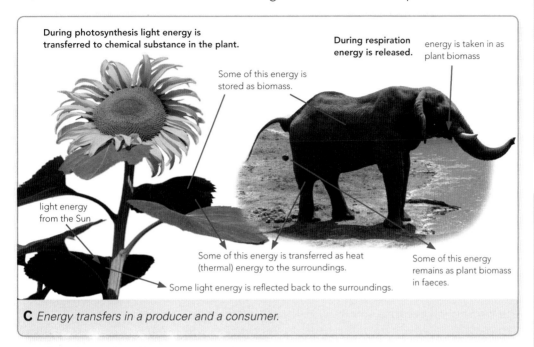

During photosynthesis light energy is transferred to chemical substance in the plant.

Some of this energy is stored as biomass.

During respiration energy is released.

energy is taken in as plant biomass

light energy from the Sun

Some of this energy is transferred as heat (thermal) energy to the surroundings.

Some light energy is reflected back to the surroundings.

Some of this energy remains as plant biomass in faeces.

C *Energy transfers in a producer and a consumer.*

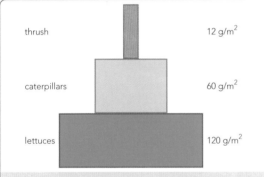

thrush 12 g/m²

caterpillars 60 g/m²

lettuces 120 g/m²

D *The width of each bar in this pyramid is a quantitative measure, showing how much biomass there is at that level.*

If we measured the biomass of all the lettuces in a field and all the caterpillars that were eating the lettuces and all the thrushes that were eating the caterpillars, we could draw a **pyramid of biomass** like the one in Figure D.

At the top trophic level of a food chain there is not enough energy in the biomass to provide enough energy for another level. This means there is a limit to the length of a food chain.

6 Describe what happens to the biomass in a food chain as you go up the trophic levels.

7 Explain in your own words why there are no predators that feed on lions.

8 Explain in terms of energy transfer how living things are interdependent.

4 Look at Figure C. List all the ways that energy leaves a food chain at the trophic level of:
a a producer
b a primary consumer.

5 Explain any differences in your answers to question **4**.

Skills spotlight

A good scientific model helps us to understand new ideas using a simplified picture that explains all of the known facts. How does the model of a pyramid of biomass help us to explain what is happening in terms of energy in a food chain? What doesn't the model show us?

Results Plus
Watch Out!

When studying a food web, remember that if one kind of animal reduces or increases in number, it is likely to affect several other organisms in different ways.

Learning Outcomes

3.16 Recall that interdependence is the dynamic relationship between all living things

3.17 Demonstrate an understanding of how some energy is transferred to less useful forms at each trophic level and this limits the length of a food chain

3.18 Demonstrate an understanding that the shape of a pyramid of biomass is determined by energy transferred at each trophic level

HSW **3** Describe how phenomena are explained using scientific models

Why do some birds ride around on large animals?

What other kinds of feeding relationships are there?

2000 years ago people thought that mistletoe (Figure A) was a 'magical' plant because it didn't have roots that grew in soil and it was green in winter. They used mistletoe in ceremonies to bring luck and love, which may explain why some kiss under it at Christmas.

A

> 1 In the human–headlouse feeding relationship, name:
> a the parasite
> b the host.
>
> 2 Explain how headlice benefit from this relationship but humans do not.

Parasites

In most feeding relationships a **predator** kills and eats its **prey** and then moves on to find more prey. **Parasitism** is a feeding relationship in which two organisims live together with one feeding off the other. The organism doing the feeding is a **parasite** and the organism it feeds on is its **host**.

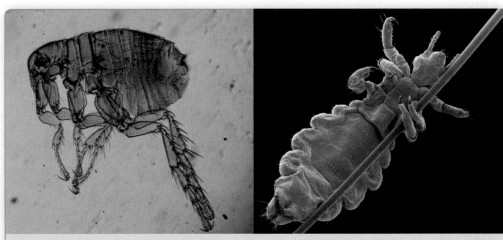

> 3 a Explain how a tapeworm can kill its host.
> b Why don't tapeworms all die out?

B *The flea (left) and the headlouse (right) are parasites. They bite other animals, such as humans, so they can feed off their blood.*

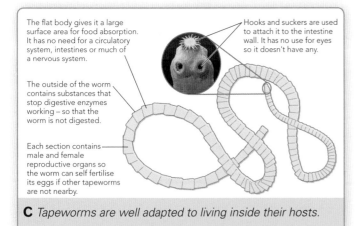

The flat body gives it a large surface area for food absorption. It has no need for a circulatory system, intestines or much of a nervous system.

Hooks and suckers are used to attach it to the intestine wall. It has no use for eyes so it doesn't have any.

The outside of the worm contains substances that stop digestive enzymes working – so that the worm is not digested.

Each section contains male and female reproductive organs so the worm can self fertilise its eggs if other tapeworms are not nearby.

C *Tapeworms are well adapted to living inside their hosts.*

Some parasites, like headlice and fleas, live on the outside of their host. Others live inside. For example, tapeworms live in vertebrate intestines. Their eggs leave the host in its faeces. Eggs swallowed by other animals hatch and grow inside their bodies.

Parasites usually harm their hosts. Tapeworms absorb nutrients from the host's gut, which can cause the host to lose a lot of weight. The worm can also grow large enough to block the host's intestines. European mistletoe is also a parasite. It has leaves that can photosynthesise but its roots grow into the veins of the host tree and absorb water and mineral salts.

Mutualists

Some organisms live in close relationships where both organisms benefit. This is called **mutualism**. Oxpeckers have a mutualistic relationship with large herbivores in Africa. Both the oxpecker and the herbivore benefit from the relationship. The oxpecker eats parasitic insects that live on the skin of the herbivore. A similar relationship is found in oceans, where small **cleaner fish** eat dead skin and parasites from the skin of larger fish, such as sharks.

D *These oxpeckers eat parasitic insects that live on the skin of this wild cow.*

Some organisms live in mutualistic relationships inside other organisms. Bacteria that can turn nitrogen in the air into nitrogen compounds are called **nitrogen-fixing bacteria**. Some live inside the roots of **legumes** (plants, like peas, that produce pods). The bacteria are protected from the environment and obtain chemical substances from the plant that they use as food. The plant gets the nitrogen compounds, which it needs to grow well, from the bacteria.

4 Explain how both the oxpecker and the wild cow benefit from their mutualistic relationship.

Chemosynthetic bacteria are producers that get their energy from chemical substances rather than from light. Some live inside giant tubeworms – the tubeworms gather the chemical substances that the bacteria need for chemosynthesis and the tubeworms feed on substances made by the bacteria.

E *Giant tubeworms live around deep-sea vents where there is no light.*

Watch Out!

It is not sufficient to state that a parasite lives on or in an organism; you must also state that the host will be harmed.

H 5 The tubeworm–bacteria relationship is mutualistic. Explain why each organism needs the other.

6 Explain fully the difference between parasitic relationships and mutualistic relationships.

Skills spotlight

Use what you know to present an argument either for or against mistletoe being classified as a parasite.

Learning Outcomes

3.19 Explain how the survival of some organisms may depend on the presence of another species:
  a parasitism, including:
    i fleas
    ii head lice
    iii tapeworms
    iv mistletoe
  b mutualism, including:
    i oxpeckers that clean other species
    ii cleaner fish
    H *iii* nitrogen-fixing bacteria in legumes
    H *iv* chemosynthetic bacteria in tube worms in deep-sea vents

HSW **11** Present information, develop an argument and draw a conclusion, using scientific, technical and mathematical language and ICT tools

B1.30 Pollution

 How do we cause pollution?

In some cities where the human population is still growing rapidly, people live in conditions where parasites and infectious diseases such as cholera can spread easily .

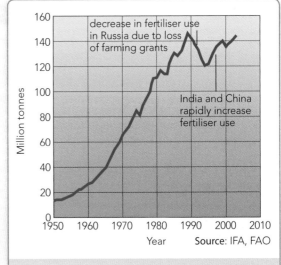

A *The population is growing so rapidly that there aren't enough houses for everyone with drains for sewage and piped clean water.*

<table>
<tr><td>? 1 Explain why the human world population has grown.</td></tr>
</table>

World human population growth

During the nineteenth and twentieth centuries there was rapid human **population growth**. Increases in food production, medicines and better living conditions for many meant that more babies were born and survived to have their own children. The rate of population growth is now slowing down but estimates that make different assumptions about how many people will die each year and how many babies will be born predict different population sizes in the future.

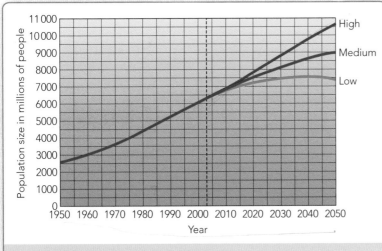

B *The UN data for human world population from 1950 to their best estimate for 2050*

C *Estimated world fertiliser use from 1950 to 2003*

Maths skills

Figures B and C are **line graphs**.

Points are plotted on the graph from pairs of observationsand then joined with straight lines.

? 2 Figure B shows three different predictions that the UN has made for human population size in 2050.
a What are the three predictions?
b Suggest why the UN made three different predictions.
c Suggest why different estimates give different predictions of human population size in the future.

3 Some estimates suggest humans use 60% of available fresh water yearly. Predict how this will change over the next 50 years.

People need food and fresh water to survive. As the human population increases, we need to find more water and produce more food. Crops often grow better with **fertiliser** (containing nitrates and phosphates) added to the soil. These compounds are nutrients needed by plants to grow well.

Everything that we use every day requires resources, including fossil fuels to generate electricity to make them.

Causing pollution

If we are not careful about the way we make and use resources, we risk releasing **pollutants** into the environment and damaging the organisms. Sulfur dioxide gas is released from burning fossil fuels and pollutes the air if the concentration is high.

If farmers use too much fertiliser, or rain washes it away, it can get into water and raise the natural concentration of nitrates and phosphates. This is called **eutrophication**.

As nutrient concentration increases, organisms in the water are affected, which can lead to a decrease in oxygen concentration in the water and the death of many animals. This is shown in Figure D below.

ResultsPlus
Watch Out!

Eutrophication is an increase of nutrients in water. The consequences of this follow in a sequence. The sequence order is important when explaining the consequences of eutrophication.

Skills spotlight

Satellites in orbit above the Earth can measure the levels of soil nutrients in a field and send the information direct to the tractor to tell the farmer exactly how much fertiliser to use. Suggest how this could help reduce water pollution.

4 Explain why fertiliser use is increasing.

5 Draw a flowchart to show how eutrophication is caused and how it can damage the environment.

6 Explain in as much detail as you can why increasing population growth and industrialisation of a country could lead to increased pollution.

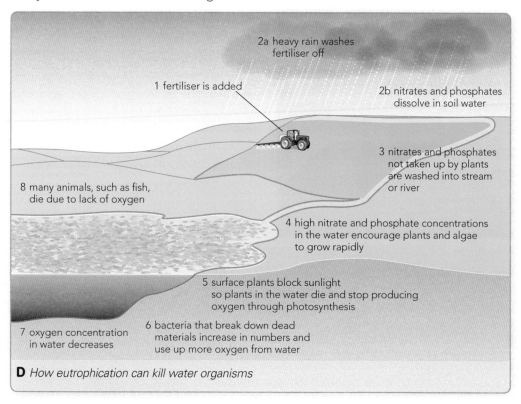

2a heavy rain washes fertiliser off

1 fertiliser is added

2b nitrates and phosphates dissolve in soil water

3 nitrates and phosphates not taken up by plants are washed into stream or river

8 many animals, such as fish, die due to lack of oxygen

4 high nitrate and phosphate concentrations in the water encourage plants and algae to grow rapidly

5 surface plants block sunlight so plants in the water die and stop producing oxygen through photosynthesis

7 oxygen concentration in water decreases

6 bacteria that break down dead materials increase in numbers and use up more oxygen from water

D *How eutrophication can kill water organisms*

Learning Outcomes

3.20 Analyse, interpret and evaluate data on global population change

3.21 Explain how the increase in human population contributes to an increase in the production of pollutants, including phosphates, nitrates and sulfur dioxide

3.22 Explain how eutrophication occurs and the problems associated with eutrophication in an aquatic environment

HSW 12 Describe the benefits, drawbacks and risks of using new scientific and technological developments

B1.31 Pollutants and plant growth

How does acid rain affect the growth of seedlings?

The 'smoke' in Figure A is a cloud of volcanic gases rising from a volcano in Nicaragua. The acid from the volcano has damaged the trees you can see in the photograph, and Figure B shows the damage done to the limestone blocks in a nearby footpath.

The gas cloud from the volcano causes damage because it contains a lot of sulfur dioxide, which dissolves in the water vapour in air to make a very strong acid.

A *The Masaya volcano in Nicaragua*

Sulfur dioxide is also produced by power stations. When it dissolves in the water vapour in clouds it forms sulfuric acid and can fall as acid rain. Acid rain is responsible for destroying the trees in many forests around the world. It can also make lakes too acidic for fish and other water organisms to live in.

B *The cement between the stone blocks is more resistant to the acid from the volcano.*

Your task

You are going to plan an investigation that will allow you to find out about the effect of acidic rainwater on plants. You are going to investigate the hypothesis that the acidity of rainwater affects how well plant seeds will germinate and grow. Your teacher will provide you with some materials to help you organise this task.

Learning Outcomes

3.23 Investigate the effect of pollutants on plant germination and growth

Build Better Answers

When completing an investigation like this, one of the skills you will be assessed on is your ability to *process your* data. There are 4 marks available for this skill. Here are two student extracts that focus on this skill. Other skills that you need for the assessment are dealt with in other lessons.

ResultsPlus

To access 2 marks
- Attempt to process all your collected evidence
- Use appropriate maths skills
- Present your processed evidence in an appropriate way

To access 4 marks
- Process all your evidence in an appropriate way, using maths skills if appropriate
- Present it in a way that allows you to draw a conclusion

Can organisms tell us about pollution?

The River Irk was polluted in April 2009 by chemicals that turned it into what looked like a giant bubble bath. This obvious pollution did look fun, but it damaged wildlife. It was linked to a spill from a local soap factory.

A

1 Explain the term 'indicator species'.

2 Why is clean air not good news for rose growers?

Air pollution indicators

The more pollution we cause, the more harm there is to habitats. Some organisms are so sensitive to polluting chemicals that we can use them to help us show the presence of pollution. We call them **indicator species**. For example, **blackspot fungus** is an infection of roses that is killed by sulfur dioxide pollution in the air.

Lichens are mutualistic relationships between a fungus and an alga that lives inside the fungus. Different species of lichen can tolerate different amounts of sulfur dioxide and other polluting gases such as nitrogen oxides and indicate the presence or absence of these gases in the air. Lichen surveys in the UK have shown that laws to reduce the amount of sulfur dioxide released into the air are working well.

B *Blackspot fungus can kill roses.*

ResultsPlus
Watch Out!

Collecting and presenting data on indicator species and then interpreting it is a valuable skill which is often tested in examinations.

C *The lichen species on the left dies quickly if there is sulfur dioxide in the air. The one on the right can grow when there is sulfur dioxide present.*

Water pollution indicators

Different animals that live in water need varying amounts of oxygen. For example, **stonefly larvae** and **freshwater shrimps** need lots of oxygen, but **bloodworms** and **sludgeworms** are adapted to live where there is little oxygen in the water. We can use the presence or absence of these organisms as simple indicators of the level of pollution in water.

D *Bloodworms have red haemoglobin to help them get dissolved oxygen from water.*

The importance of recycling

In the UK we produce over 100 million tonnes of waste each year from homes, businesses and industry. This waste includes paper, plastics and many metals, and much of it ends up buried in the ground in landfill sites. This not only uses a lot of land, there is also a risk of pollution and it means that the materials cannot be used again. We are in danger of needing more of some materials than we can supply and even of running out of some raw materials completely.

Metal	Use	Predicted years before we need more than we can supply
gallium	flat screen TVs	c.15
platinum	catalytic converters	c.15
zinc	cars, roofs, computers	c.30

E *The reserves of some metal elements in the ground may run out soon.*

3 Name two indicator species of clean water and two of polluted water.

4 a Why do animals need oxygen?
b Suggest why plants are not used as indicators of water pollution.

Recycling is the process of taking materials out of waste before disposal and converting them into new products that we can use:

- Metals in drinks cans can be melted down and recycled as new drinks cans or part of a car.
- Paper can be recycled as more paper or cardboard.
- Plastic bottles can be recycled as fleece clothing. Recycling can help supply more of the materials we need that are difficult to get hold of or are running out.

5 Suggest why running out of gallium would be a problem.

6 Describe fully the advantages of recycling.

Skills spotlight

How could you use the presence/absence (qualitative data) or number (quantitative data) of different lichen species to indicate the level of sulfur dioxide pollution?

Learning Outcomes

3.24 Demonstrate an understanding of how scientists can use the presence or absence of indicator species as evidence to assess the level of pollution:
   a polluted water indicator – bloodworm, sludgeworm
   b clean water indicator – stonefly, freshwater shrimps
   c air quality indicator – lichen species, blackspot fungus on roses

3.25 Demonstrate an understanding of how recycling can reduce the demand for resources and the problem of waste disposal, including paper, plastics and metals

HSW 10 Use qualitative and quantitative approaches when presenting scientific ideas and arguments, and recording observations

The carbon cycle

 How do natural processes use carbon?

If it weren't for billions of natural recycling organisms, the Earth would be covered in dead bodies and waste.

A *A fungus slowly digests the dead body of a weevil from the inside out.*

> **1** Which process removes carbon dioxide from the air?
>
> **2** Describe and compare how carbon dioxide is used in photosynthesis and respiration.

Capturing carbon dioxide

Imagine a carbon atom that is part of a carbon dioxide molecule in the atmosphere. That carbon dioxide molecule could diffuse into a leaf. There, during photosynthesis, the carbon atom from the carbon dioxide molecule will become part of another carbon compound called glucose.

If the glucose is used by the plant for respiration, the carbon atom will become part of carbon dioxide again and will be released back into the atmosphere. However, the glucose might instead be changed into other carbon compounds and used to make more plant biomass.

Carbon compounds in food chains

> **3** Draw a diagram to show what happens to the carbon in an animal that is eaten by a predator.

Carbohydrates, fats and proteins in plants are all compounds that contain carbon atoms. When an animal eats a plant, some of these compounds are broken down during digestion and taken into its body. The rest will leave the animal's body as **faeces**.

Some of the carbon compounds taken into the body will be used for respiration and some may form waste products that are excreted in urine. The rest are used to build more complex compounds inside the animal and make more animal biomass.

If the animal is eaten by a predator, the same processes will happen again.

If plants and animals are not eaten and just die, detritus feeders, such as worms and fly larvae (maggots) and **decomposer** organisms, such as fungi and bacteria, start the process of **decay**. They also feed on animal waste (e.g. faeces, urine). They break down the carbon-containing compounds and use some of them for respiration and the rest to build more complex compounds in their bodies. When they die, their bodies will be broken down by other detritus feeders and decomposers and so on.

B *Detritus feeders are often the first stage in decay of a dead body.*

Beyond the food chain

If many large dead plants are buried so quickly that decomposer organisms can't feed on them, then over millions of years they may be changed by heat and pressure into peat or coal. In the same way, oil and natural gas are formed from tiny dead sea plants and animals which have not been decomposed.

Coal, peat, oil and natural gas are **fossil fuels**; they contain carbon compounds formed from dead organisms. When we burn fossil fuels, in the process called **combustion**, the carbon in these compounds is converted to carbon in carbon dioxide and is released into the atmosphere.

The movement of carbon through dead and living organisms and the atmosphere is shown in the **carbon cycle** and is an example of natural recycling.

Results Plus
Watch Out!

Photosynthesis is a process that involves the build up of molecules to create glucose. Respiration is the process that involves the breakdown of glucose to release simple molecules, including carbon dioxide.

Skills spotlight

Explain how the carbon cycle can help scientists identify possible causes of increasing amounts of carbon dioxide in the atmosphere.

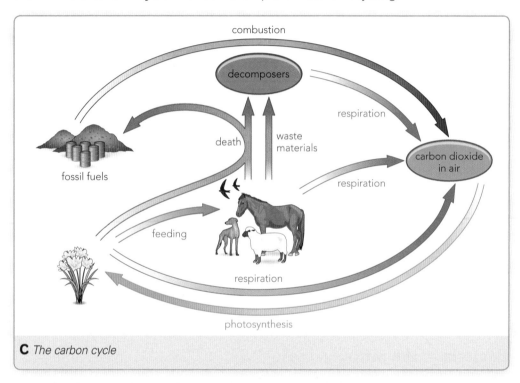

C *The carbon cycle*

4 Which processes release carbon dioxide into the atmosphere?

5 The rate of photosynthesis depends partly on temperature. Using the carbon cycle model, suggest how global warming could affect the amount of carbon dioxide in the atmosphere.

Learning Outcomes

3.26 Demonstrate an understanding of how carbon is recycled:
  a during photosynthesis plants remove carbon dioxide from the atmosphere
  b carbon compounds pass along a food chain
  c during respiration organisms release carbon dioxide into the atmosphere
  d decomposers release carbon dioxide into the atmosphere
  e combustion of fossil fuels releases carbon dioxide into the atmosphere

HSW **3** Describe how phenomena are explained using scientific models

The nitrogen cycle

How does nitrogen cycle through the environment?

This salad is packaged in 100% nitrogen gas. As most decomposer organisms need oxygen for respiration, this stops them growing and keeps the food fresher for longer.

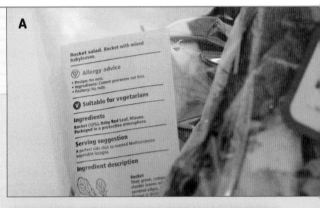

A

1 Which of the following found in animals contain nitrogen:
muscle
enzymes
hormones
brain tissue
blood?

2 Using what you have learned about carbon, draw a diagram to show what happens to the nitrogen in one animal that is eaten by a predator. (*Hint*: nitrogen is lost in urine too.)

3 Explain why decomposers are important in the cycling of nitrogen through the food web.

Nitrogen in plants and animals

Nitrogen is important for organisms because it is used to make proteins and other substances in new tissues as they grow. However, plants and animals cannot use nitrogen gas from the air; they need to get nitrogen from nitrogen-containing substances. Plant roots absorb nitrogen compounds, such as nitrates, from the soil and use them to make proteins and other nitrogen compounds in their biomass. Animals get nitrogen compounds in their food. They use most of these to make nitrogen compounds in new biomass, but some are lost to the environment in faeces and urine.

B *If nitrogen is in short supply an organism cannot grow well. The plant on the left has been deprived of nitrogen.*

Bacteria in the nitrogen cycle

If there were no way to convert nitrogen compounds from biomass and waste back to nitrates in the soil, plants would quickly die out. Fortunately, when decomposers feed on dead plants and animals and animal wastes they break down some of the proteins and **urea** (a nitrogen-rich substance in urine) to ammonia and release it into the soil.

Other bacteria also play a part in keeping the soil healthy for plant growth. The soil contains **nitrogen-fixing bacteria** that can fix nitrogen gas into ammonia.

Mutualistic nitrogen-fixing bacteria are also found in legume roots. These bacteria live inside **root nodules** and provide these plants directly with ammonia.

Although plants can use ammonia, they grow better with nitrates. Some **nitrifying bacteria** in the soil convert ammonia to nitrates.

C *This pea plant has made special root nodule structures for nitrogen-fixing bacteria to live in.*

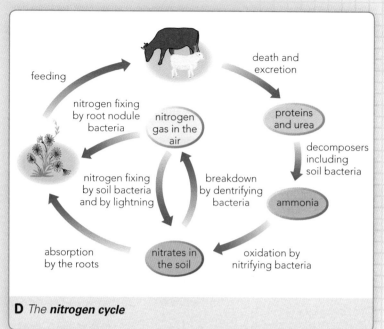

D *The **nitrogen cycle***

If soils are lacking in oxygen, such as when they are waterlogged, then some **denitrifying bacteria** will convert the nitrates back to nitrites and others convert nitrites back to nitrogen gas.

Occasionally, lightning can provide the energy to combine oxygen and nitrogen gases in the air, forming oxides that quickly form nitrates.

Results Plus
Watch Out!

The nitrogen cycle contains nitrogen - fixing bacteria, nitrifying bacteria, denitrifying bacteria and decomposers. Try to be clear about the roles of each type.

Skills spotlight

Present the results shown in Figure E in a suitable graph. Use the graph to help you draw a conclusion about the effects of root nodule bacteria on amount of seed produced.

Area	Amount of seed produced (kg/ha)		
	Control	+ fertiliser	+ root nodule bacteria
A	2341	2432	2462
B	2483	2660	3119

E *Results of an investigation into the effect of using either nitrogen fertiliser or root nodule bacteria on the amount of seed produced by a legume crop (soybean).*

4 Draw a labelled diagram to show how nitrogen cycles through food chains, including producers, primary consumers, secondary consumers and decomposers.

5 Explain why adding manure to the soil increases the nitrate content of the soil.

6 Describe the stages in which soil bacteria change the amounts of plant nutrients in the soil.

Learning Outcomes

H 3.27 Demonstrate an understanding of how nitrogen is recycled:
  a nitrogen gas in the air cannot be used directly by plants and animals
  b nitrogen-fixing bacteria living in root nodules or the soil can fix nitrogen gas
  c the action of lightning can convert nitrogen gas into nitrates
  d decomposers break down dead animals and plants
  e soil bacteria convert proteins and urea into ammonia
  f nitrifying bacteria convert this ammonia to nitrates
  g plants absorb nitrates from the soil
  h nitrates are needed by plants to make proteins for growth
  i nitrogen compounds pass along a food chain or web
  j denitrifying bacteria convert nitrates to nitrogen gas

HSW 11 Present information, develop an argument and draw a conclusion, using scientific, technical and mathematical language and ICT tools

These questions are indicative of the type of questions used in the exam. Refer to page 6 for information on the grades.

Bird classification

1. The photographs show three birds. Each bird has a common name and a scientific name.

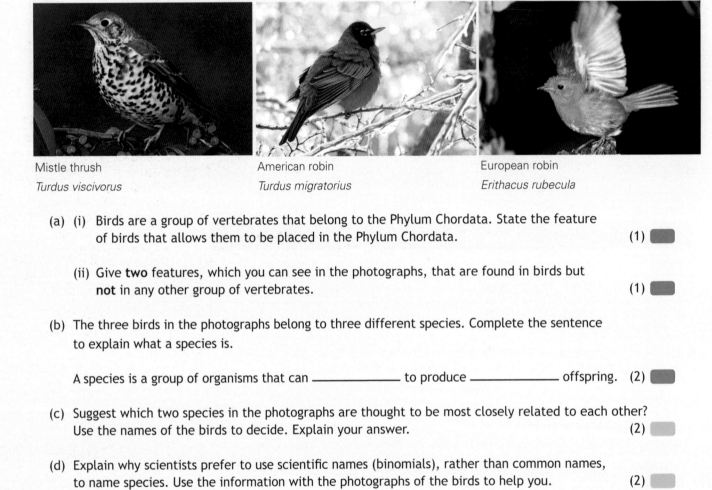

Mistle thrush

Turdus viscivorus

American robin

Turdus migratorius

European robin

Erithacus rubecula

(a) (i) Birds are a group of vertebrates that belong to the Phylum Chordata. State the feature of birds that allows them to be placed in the Phylum Chordata. (1)

(ii) Give **two** features, which you can see in the photographs, that are found in birds but **not** in any other group of vertebrates. (1)

(b) The three birds in the photographs belong to three different species. Complete the sentence to explain what a species is.

A species is a group of organisms that can _____ to produce _____ offspring. (2)

(c) Suggest which two species in the photographs are thought to be most closely related to each other? Use the names of the birds to decide. Explain your answer. (2)

(d) Explain why scientists prefer to use scientific names (binomials), rather than common names, to name species. Use the information with the photographs of the birds to help you. (2)

Food chains in Africa

2. Buffalo live on the African savannah. Small animals called ticks feed on the buffalo. Birds called oxpeckers feed on the ticks.

(a) (i) The ticks feed on the buffalo and can make the buffalo ill. Choose the correct term for the relationship between the ticks and the buffalo.

    A eutrophication
    B mutualism
    C parasitism
    D phototropism (1)

(ii) The oxpeckers remove the ticks, which helps the buffalo to stay healthy. Choose the correct term for the relationship between the oxpecker and the buffalo.

    A eutrophication
    B mutualism
    C parasitism
    D phototropism (1)

(b) The food chain shows another set of feeding relationships between organisms on the African savannah.

grass   →   buffalo   →   hunting dogs

The diagram shows a pyramid of biomass for this food chain.

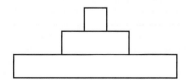

(i) Copy the diagram and write the name of each organism in the food chain next to the box that represents it. (2)

(ii) Pyramids of biomass for other food chains are often the same shape as the one shown in (i). Explain why pyramids of biomass are usually this shape. (3)

Houseflies and dysentery

3. Dysentery is caused by bacteria. The bacteria are often transmitted to food or drink by houseflies.

(a) (i) What is the correct term for an organism, such as the dysentery bacterium, that causes a disease?

    A   infection
    B   pathogen
    C   vector
    D   virus   (1)

  (ii) What is the correct term for an organism, such as a housefly, that spreads a disease-causing organism?

    A   infection
    B   pathogen
    C   vector
    D   virus   (1)

(b) Name one disease, other than dysentery, that is spread by insects, and name the insect that spreads this disease.   (2)

(c) An experiment was carried out to find out whether houseflies spread dysentery. The experiment took place in two military camps, where thousands of soldiers were living. Houseflies were killed at camp **A** but not at camp **B**. The table shows the results.

	Camp **A**	Camp **B**
Were houseflies killed?	Yes	No
Percentage of soldiers who got dysentery	3	12

(i) Suggest an explanation for these results.   (2)

(ii) Explain what the researchers could have done to improve the quality of their data.   (3)

Plant hormones

4. Pete did an experiment to investigate how roots respond to gravity. He took two bean seedlings and placed each seedling on its side. He put both seedlings into a dark place and left them for one day. The diagram shows the appearance of the seedlings at the start of the experiment, and after one day.

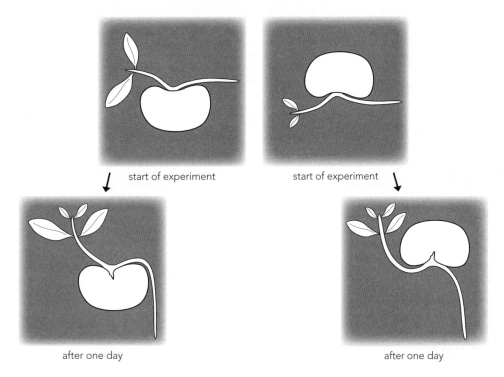

start of experiment

start of experiment

after one day

after one day

(a) (i) What do these results show?

    A   Roots grow towards gravity.
    B   Roots grow towards light.
    C   Shoots grow towards gravity.
    D   Shoots grow towards light.   (1)

(ii) What is the name for the response shown by the roots of the seedlings?

    A   negative geotropism
    B   negative phototropism
    C   positive geotropism
    D   positive phototropism   (1)

(b) Pete was investigating the response of roots to gravity. Explain why it was a good idea for Pete to put the seedlings in a dark place, rather than in the light.   (2)

(c) Explain why it is useful to the plant for its root to grow downwards.   (2)

(d) A plant hormone helps shoots to respond to light. Name the hormone and explain how it makes shoots grow towards the light.   (6)

Obesity, diabetes and heart disease

5. In a healthy person, the level of glucose in the blood is kept fairly constant. One of the ways the body does this is by secreting insulin from the pancreas.

blood glucose level rises
↓
pancreas increases secretion of insulin
and decreases secretion of glucagon
↓
blood glucose level falls

(a) (i) Choose the correct word to complete the sentence.
High blood glucose level is reduced by insulin, which is ——————————— .

    A an enzyme
    B a hormone
    C a receptor
    D a sugar

(1)

(ii) In which organ is glucose converted to glycogen?

    A brain
    B liver
    C stomach
    D pancreas

(1)

(b) (i) Diabetes is a condition in which the body cannot control the blood glucose level.
Complete the sentences to explain the causes of Type 1 diabetes and Type 2 diabetes.

Type 1 diabetes is caused by ——————————— .
Type 2 diabetes is caused by ——————————— .

(2)

(ii) State **two** ways in which a person with Type 2 diabetes can control their blood glucose level.

(2)

(c) Obesity also increases the risk of developing heart disease. Heart transplants are very expensive. The National Health Service only has a limited amount of money that it can use for transplants.

The graph shows how a man's body weight affects the chances of survival for at least five years after a heart transplant.

Discuss whether people who have normal body weight should be given heart transplants in preference to obese people. You should refer to the information above and the graph in your answer. (6)

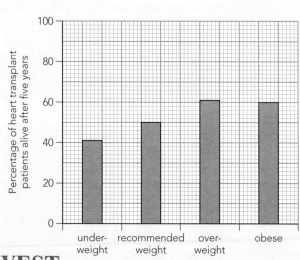

Dalmatian dogs

1. Dalmatian dogs can have brown spots or black spots. The colour of the spots is determined by a gene that has two alleles, B and b. Allele B gives black spots. Allele b is recessive and gives brown spots.

(a) (i) What is the phenotype of a dalmatian dog with the genotype BB? (1) ▢

(ii) What is the genotype of a dalmatian dog heterozygous for spot colour? (1) ▢

(b) A breeder has a dalmatian dog with black spots. She wants to know if it has the genotype BB, or Bb. She decides to breed her black-spotted dalmatian dog with a brown-spotted dalmatian dog. Copy and complete the Punnett square to show the likely results of the cross if the black-spotted dalmatian dog has the genotype Bb.

	gametes from black-spotted dalmatian dog	
	B	b
gametes from brown-spotted dalmatian dog		

(2) ▢

(c) The breeder crosses two black-spotted dalmatian dogs that each have the genotype Bb. Twelve puppies are born. Use a Punnett square to help you to predict the probable number of puppies with black spots and brown spots. (3) ▢

(d) There are many other varieties of domestic dogs, which differ greatly from dalmatian dogs. Give **one** reason why all domestic dogs are classified as belonging to the same species. (1) ▢

Marathon runner

2. When an athlete runs, respiration in the muscle cells generates heat energy. This increases body temperature. The graph shows how the body temperature of a marathon runner changed during a race and after he stopped running.

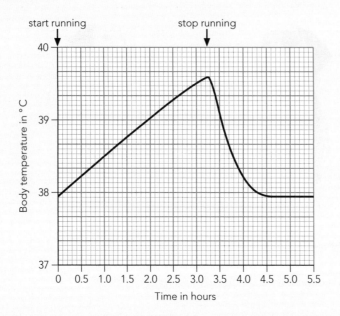

Source: Journal of the American College of Cardiology (JACC)

(a) (i) How long did it take for the runner's body temperature to return to normal after the race finished? (1)

    (ii) What was the increase in the body temperature of the runner during the race? (1)

(b) Explain why a marathon runner should drink plenty of fluid during the race. (2)

(c) Explain how the blood vessels in the runner's skin will help him to cool down after the race. (3)

Caffeine and reaction time

3. Angie investigated how drinks containing caffeine affect a person's reaction time. Angie found a website to measure reaction time. She clicked on the start button. When the traffic light turned green she clicked on the button again.

Click here to start

Angie tried the test before she had a drink containing caffeine. She had five tries. Then she drank a large cup of strong coffee and did the test again. The table shows her results.

try	reaction time in ms	
	before drinking coffee	after drinking coffee
1	279	240
2	251	238
3	246	237
4	243	232
5	241	229
mean	252	

(a) (i) Calculate Angie's mean reaction time, after drinking coffee. (1)

(ii) Suggest why it was a good idea to have five tries, rather than just one. (1)

(iii) What was Angie's fastest response time before drinking coffee?

A 241 ms
B 243 ms
C 246 ms
D 251 ms (1)

(iv) Angie made a conclusion. She said that drinking coffee reduces reaction time. Jo said that her results did not really support this conclusion. State **one** reason why Jo is correct, using Angie's results to help you. (1)

(b) When Angie was doing the reaction time test, information was picked up by receptors in her eyes. This information was transferred to the muscles in her finger as electrical impulses along neurones. The information passed through several structures in Angie's body.

Write the numbers 2 to 4 next to the structures, to complete the sequence in which the information travelled. Numbers 1 and 5 have already been completed.

motor neurone	
sensory neurone	
receptors in eye	1
muscles in finger	5
central nervous system	

(2)

(c) Describe how the electrical impulse is transmitted between two neurones. (3)

Natural selection in lizards

4. The lizard *Anolis sagrei* lives on several islands in the Caribbean. It spends part of its time on the ground, and part of its time on branches of bushes and trees.

(a) (i) State the name of the group of vertebrates to which the lizard belongs. (1)

(ii) Give one piece of evidence from the photograph to support your answer to (i). (1)

The lizards get onto branches by jumping. Scientists investigated this hypothesis:

The longer the length of a lizard's legs, the higher it can jump.

First, they measured the length of a lizard's legs. They then placed the lizard on a flat surface and clapped their hands to make it jump. They did this five times for each lizard and measured the height of the highest jump each lizard made. Their results are shown in the graph.

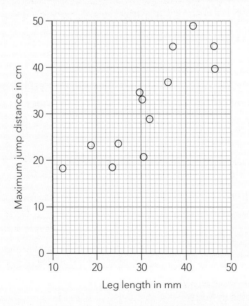

Source: The Company of Biologists Ltd

(b) (i) Suggest **two** variables that the scientists should have kept constant during their experiment. (2)

(ii) State whether or not these results support the researchers' hypothesis. Explain your answer. (2)

(c) The scientists measured the leg lengths of the lizards on two islands where there were no predators. They found that the lizards on both islands had similar leg lengths. The scientists introduced a predator, the curly-tailed lizard, onto one of the islands. The predator killed *Anolis sagrei* lizards. The predator hunted on the ground, not up in the branches.

The scientists found that, after several years, the mean leg length of the *Anolis* lizards on the island with the predator increased. The mean leg length of the *Anolis* lizards on the island with no predator did not change. Suggest an explanation for the increase in leg length of the lizards on the island with the predator.

(6)

Fertilisers

5. Farmers often add fertilisers to fields where they are growing crops such as wheat. The fertilisers contain nitrate ions, which can increase the growth of the wheat plants, so the farmer harvests a larger quantity of grain.

(a) (i) Name the part of the plant that absorbs the nitrate ions. (1)

   (ii) Name one type of substance that the plant can make, using the nitrate ions. (1)

(b) One type of bacteria in the soil breaks down the nitrate ions, so that the nitrogen is not available to the plants. Name this group of bacteria, and state the substance that they produce from the nitrate ions. (2)

(c) Some crop plants, such as beans, have nodules on their roots. Explain why a farmer does not need to add fertilisers containing nitrate ions to these crops. (2)

(d) When the wheat grain is harvested, the dead stems of the wheat plants are often ploughed back into the soil. Explain how different types of organisms living in the soil can use the dead plant material to produce nitrate ions. (6)

Here are three student answers to the following question. Read the answers together with the examiner comments around and after them.

Question	Light and plant growth		Grade	G–D

Claire carried out an investigation to find out how light affected the growth of plants.

She placed 8 cress seeds into each of the dishes A, B, C and D. Claire shone light onto dishes A, B and C from different directions. Dish D was kept in the dark. Claire left the dishes for one week. The diagrams show the results of Claire's investigation.

Claire controlled several variables when she carried out her investigation. Describe which variables Claire could control, how she could control these variables and why they need to be controlled. (6)

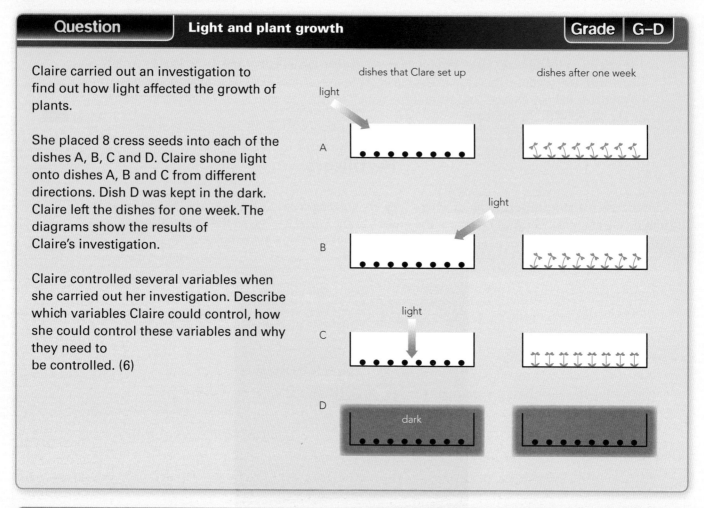

dishes that Clare set up   dishes after one week

Student answer 1   Extract typical of a level ① answer

Good – two important variables have been mentioned.

She should make sure they all have the same amount of water and temperature. She could measure the water in a measuring cylinder so they all get the same amount. They should all be watered with the same amount of water at the same time every day.

That's good - it states how the water would be measured.

Examiner summary
Everything in this answer is correct, but there are many more variables that should have been controlled. The answer only includes water and temperature and explains how Claire could control the amount of water. However, the answer could also explain *why* any of the variables need to be controlled.

Good – that is an important variable to be controlled.

Two more important variables mentioned.

> • She needs to have the same number of seeds in each dish.
> • All the seeds should be the same kind and out of the same packet. She must water all the seeds and keep them in a place that is kept at the same temperature. This is because seeds need water and warmth to germinate. If there were too many seeds they might not be able to grow properly in the dish.

This explains why the seeds need water and warmth, but not why these should be controlled.

Examiner summary

This answer includes several different variables, which is good. It says that the seeds should be from the same packet, but it needs to explain *why* this is important – for example, that the seeds should be the same age or have been kept in the same conditions.

A good list of five important variables that should be controlled.

This explains how each of the variables should be controlled.

> She needs to control water, temperature, number of seeds, age of seeds and type of seeds. Measure water in a measuring cylinder and pour the same volume into each dish each day. Put them all in the same place so the temperature is the same, and check using a thermometer. Put exactly 10 seeds in each dish, making sure they are all the same age and same kind. This is because we want to know how light affects the seeds, so everything else should be kept the same for all of them. If some of the seeds were too old, they might not germinate or grow properly.

An explanation of why the variables should be controlled has been included.

Examiner summary

This answer gives five important variables. It also describes how they should be controlled and explains why this is important.

ResultsPlus

Move from level ❶ to level ❷

Try to think of more than one or two ideas for variables, rather than spending most of your time and words on just one. Make sure you answer all the parts of the question – student 1 hasn't tried to answer why the variables should be controlled.

Move from level ❷ to level ❸

Get really organised with your answer. For a question like this one, start by making a list of four or five variables. Then look at each thing that the question asks you to do in turn and make sure your answer deals with each of them. Often, the last thing is the most difficult, but even if you aren't very confident about it, it is always worth having a go at answering it.

Here are three student answers to the following question. Read the answers together with the examiner comments around and after them.

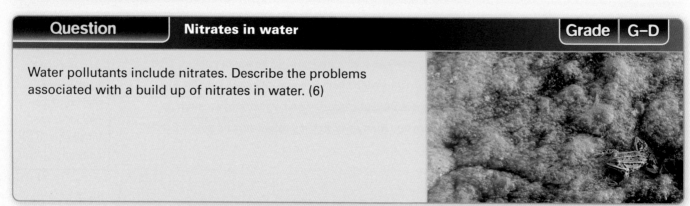

Question | **Nitrates in water** | **Grade** | **G–D**

Water pollutants include nitrates. Describe the problems associated with a build up of nitrates in water. (6)

Student answer 1 | **Extract typical of a level ① answer**

This isn't correct. The nitrates make plants grow faster.

• The nitrates are deadly poisonous and kill all the plants and animals in the water. The nitrates pollute the water and block out all the sunlight. Everything rots and make the water slimy and smelly so nothing can live in it.

'Everything' is not a good word to use. Nor is rot. It would be better to say that dead plants are decomposed by bacteria.

It's true that sunlight is blocked out, but this isn't done by the nitrates.

Examiner summary

There are a few correct ideas about how the nitrates affect the water, but the answer could include more detail here. The scientific terminology is used incorrectly because nitrates are not poisonous. The answer misses out one very important step in the process – the rapid growth of plants and algae.

Student answer 2 | **Extract typical of a level ② answer**

This is the correct scientific term for the process.

• Eutrophication happens meaning the nitrates kill the plants and they get eaten by bacteria that use up all the oxygen in the water so the fish die so they rot and make the water smell unpleasant and animals can't live in it.

This isn't correct – the nitrates make the plants grow faster.

Eaten is not a good word to use here – decomposed is the correct scientific term.

Try to use good punctuation in your answer to make it easy to read. Writing this as two sentences instead of one would be better.

Examiner summary

This answer shows understanding of some of the sequence of events that happens when nitrates pollute water. However, the scientific language could be improved by using decomposed instead of eaten. Some important points have been missed, for example why the plants die or why oxygen is used up. The meaning is not always clear – for example, the first they could either refer to the nitrates (which would be incorrect) or the plants. The answer has no punctuation, which makes it difficult to understand. It would be much better to split it up into separate sentences.

This is a clear explanation of how the nitrates cause plants to die.

The nitrates make algae and plants grow faster, so they cover the water surface and stop light getting in. This means the plants lower down in the water can't photosynthesise, so they die.

The dead plants are decomposed by bacteria. The bacteria reproduce quickly and produce a big population because they have a lot of nutrients (from the dead plants). The bacteria respire and use up oxygen, so there isn't any left for the fish and other animals and they die. This is called eutrophication.

Decompose is a good scientific term to use here.

This explains clearly why the oxygen in the water is used up.

Examiner summary

This answer clearly and correctly describes the sequence of events that takes place. There is good use of scientific language – photosynthesise, decomposed, population, nutrients, respire and eutrophication. The spelling, punctuation and grammar are all correct and the answer is easy to follow.

 Results Plus

Move from level ① to level ②

Make sure that you always use some scientific terms in your answer. If you are describing a sequence, start at the beginning and work steadily to the end, step by step – it's a good idea to make a list of the steps on a piece of rough paper before you start writing your answer on the exam paper.

Try to explain how one step causes the next one to happen. If you are thinking about that, it might help you to get the facts right if you are not quite sure of them.

Move from level ② to level ③

Use as many relevant, correct scientific terms as you can – check through your answer to see if you have used everyday language that could be changed into scientific language.

Read through your answer – would someone else reading it be certain what you mean? Try not to use 'it' and 'they', because the examiner may not be absolutely sure what you are referring to. It is usually better to use several short, clearly written sentences rather than one or two very long ones.

Here are three student answers to the following question. Read the answers together with the examiner comments around and after them.

Question	Phototropism and geotropism	Grade	E–A*

Describe how phototropism and geotropism affect the development of plants so they can grow more successfully. (6)

Student answer 1 — Extract typical of a level ① answer

You could improve this by using the word phototropism and saying which part grows towards the light.

Plants grow towards the light so they can make food better because they need light to make food. So the plant makes more food and it can grow better. The food gives it energy so it can grow. If the plant didn't grow towards the light, it wouldn't make enough food so it wouldn't grow successfully.

It would be better to use the word photosynthesis here.

Examiner summary
The answer shows some understanding of phototropism, but it doesn't include very many correct facts. The last sentence only repeats information that has already been given in the first two. The answer could be improved by mentioning geotropism.

Student answer 2 — Extract typical of a level ② answer

This starts to explain *how* phototropism happens.

Phototropism is when you have a plant with light coming from one side and the top of the plant grows towards it. This is because the shady side grows more than the light side so the plant grows towards the light so it can do more photosynthesis. Geotropism is the opposite, so it grows down into the ground so it can get water which it needs to grow properly.

This explains *why* phototropism in shoots is useful to the plant.

This is a bit muddled – you need to explain which part, i.e. the roots, grows down into the ground.

Examiner summary
This answer shows quite a bit of knowledge about phototropism and geotropism, and tries to explain how these two responses happen and why they are important to the plant. Using scientific terminology (other than the words that were already in the question) and giving more detail would improve this answer.

This explains clearly how shoots grow towards the light.

Phototropism means growing towards or away from the light. Shoots grow towards light, because auxins collect on the shady side, so it gets longer more than the light side, so the shoot bends over. This is good because it means the leaves get more light for photosynthesis, so they can make more starch which they need for energy, or for making new cellulose cell walls. This helps the plant to grow successfully because it helps it to make new cells.

Geotropism means growing towards or away from gravity. Roots grow towards gravity, so they go down into the soil where they can get water and minerals and help to hold the plant up. Water is needed for photosynthesis.

An explanation of how phototropism helps plants to grow successfully.

A good description of what geotropism means, and how this helps the plant to grow successfully.

Examiner summary

This answer is well organised and easy to follow. It deals with phototropism first and describes clearly what it is and how the shoot grows towards the light. Then it explains how this helps the plant. The same ideas are then covered for geotropism.

 Results**Plus**

Move from level ❶ to level ❷

Check that you have answered everything the question asks. Read through your answer – have you just said the same thing in two different ways? If so, cross out the repetitive part, and try to write something new instead.

Move from level ❷ to level ❸

Make sure your answer is well-organised. A good way to do this is to list the main points you are going to write about on some rough paper and put them into a sensible sequence. It is also important to use the correct specialist terms, such as auxins and cellulose.

Here are three student answers to the following question. Read the answers together with the examiner comments around and after them.

| Question | Nitrogen recycling | Grade | D–A* |

Explain how nitrogen is recycled in the environment. You should include the role of the different soil bacteria in your answer. (6)

Student answer 1 · Extract typical of a level ① answer

This is not correct as nitrogen can easily be *absorbed* into a plant, but the plant cannot *use* it for anything.

Nitrogen can't be absorbed by plants, so there are soil bacteria that do this instead. The bacteria absorb the nitrogen and give it to the plants. Then animals eat the plants and get their nitrogen. Other bacteria make the nitrogen go back into the air again.

Again, you need to name the bacteria, and explain what they do.

You should give the name of the bacteria, and the form in which nitrogen is taken up and used by plants.

Examiner summary

This answer includes some incorrect facts and doesn't show much correct knowledge about the nitrogen cycle. However, it tries to describe the sequence of events, where nitrogen moves from the air, then into bacteria, then into plants and then into animals. To improve this answer, a lot more detail is needed. It needs to include all important steps, for example, what happens to nitrogen in dead animals or their excretory products.

Student answer 2 · Extract typical of a level ② answer

It is good that the nitrogen-fixing bacteria are named, but how do they help the nitrogen to go into plants?

Nitrogen in the environment goes from the air into the soil where there are nitrogen-fixing bacteria that help the nitrogen to go into plants. The plants can use the nitrogen to make proteins.
Animals get their nitrogen by eating plants. Bacteria break down dead animals to put more nitrogen into the soil. They are called denitrifying bacteria. There are also nitrifying bacteria which complete the nitrogen cycle. Then it all goes round again.

This makes the form of the nitrogen in plants clear.

Make sure you use scientific terms correctly – denitrifying and nitrifying are wrongly used here.

Examiner summary

This answer gives some of the steps in the movement of nitrogen between the air, the soil, plants and animals, and they are in a sensible sequence. Several kinds of bacteria are named, but the answer does not always say what they do. It is good that the bacteria have been named, but unfortunately two of the names are wrong. In general, this answer needs more detail.

Student answer 3 | **Extract typical of a level ③ answer**

Good - this explains the role of nitrogen-fixing bacteria.

Nitrogen gas in the air is unreactive so it can't be used by plants. Nitrogen-fixing bacteria change it into ammonium ions, which plants can use to make proteins. These bacteria are often in root nodules, or sometimes they live in the soil.

It would have been good to use the word decomposers here.

Other bacteria break down urea (from animals) and dead plants and animals to make ammonia. Nitrifying bacteria change the ammonia into nitrates in the soil, which plants can take up through their roots and use them to make proteins. Animals get nitrogen from proteins in plants and other animals that they eat.

Denitrifying bacteria change nitrates in the soil into nitrogen gas, which goes back into the air.

This is a clear explanation of how the cycle is completed.

Examiner summary

This is a very good answer because it has plenty of correct and relevant details, uses scientific terminology correctly and is easy to follow. It describes the steps in a clear sequence, explaining how one step leads into the next. It states the form in which the nitrogen is found at different stages of the cycle, and names the bacteria that carry out the different steps.

 Results**Plus**

Move from level ① to level ②

Always try to use scientific language, and use it correctly. Check through your answer, and make sure you have used a scientific term wherever possible.

Make sure your answer contains some correct details. If the question asks you to Explain, then make sure you say *how* and *why* something happens, rather than just *what* happens.

Move from level ② to level ③

Use plenty of scientific language – but do make sure you get it right! Include as much detail in your answer as you can.

When you are describing a sequence, check that you haven't missed out any important steps. Each step described in your answer should link to the next one.

Explain questions are a bit more difficult than Describe ones, because you need to show that you understand how and why things happen.

Be the Examiner

Exam question report

Smokers are more likely than non-smokers to get lung infections. These lung infections are caused by microbes that are carried: (1)

A in blood　　　　**B** in food　　　　**C** by mosquitoes　　　　**D** in air

Answer: The correct answer is D.

How students answered

More than half of the students who answered this question thought that the answer was A rather than D. Students who got this right knew that smoking damages the lungs and makes it easier for microbes in the air to get into the lungs and cause infection.

A broken pump allowed 100 million litres of raw sewage to flow into a Scottish river. Scientists measured the number of algae per cm³ of water from the river for the first 20 days after the sewage spillage, and compared their data to average numbers from the same river for the previous ten years. The measurements showed that the number of algae increased for the first ten days after the sewage entered the river.

a **Explain what caused the increase of algae for the first ten days after the sewage entered the river. (3)**

 Correct answer: Any three points from: eutrophication; extra nitrates in the water; extra phosphates in the water, which are used to make plant proteins; or sewage has killed the animals that eat the algae.

Many students just gave one point and so got one mark. Only a few students gave three points and got three marks.

b **Suggest one reason why the number of algae had returned to average figures within 20 days of the sewage spill. (1)**

Correct answer: There were three possible answers that would have gained this mark:
- The sewage had flowed down the river away from the location where the algae were being measured.
- The nitrates (or phosphates) in the sewage had been used up by the growing algae.
- The extra algae had been eaten by herbivores.

Well over half of students got no marks for part b. Just over a third of students made one of those three points for the mark.

Exam question report

A survey found out that 19.8% of overweight or obese children had parents who were either overweight or obese. What can be concluded from these data? (1)

A that obesity is inherited　　**B** that obesity is inherited but there is an environmental influence　　**C** that obesity is influenced only by the environment　　**D** no definite conclusion can be made

Answer: The correct answer is D.

How students answered

　　　　　0 marks　　　　　　　　　　1 mark

Almost half of wrong answers were for option B and just under a quarter of students chose option C.

Many students don't like to choose an option like D because they assume they wouldn't be asked a question without a definite conclusion. But sometimes the correct answer really is that you can't draw a conclusion from the data you've been given.

Exam question report

Which is an example of natural selection? (1)

A the cultivation of flowers with a particular colour   **B** putting DNA into bacteria   **C** the development of long necks in giraffes   **D** breeding cows that produce lots of milk

Answer: The correct answer is C.

How students answered

0 marks

Approximately three quarters of students picked one of the wrong answers.

1 mark

Only about a quarter of students worked out that cultivation is a human activity so it is not an example of natural selection. Putting DNA into bacteria is genetic modification. Breeding cows that produce lots of milk is selective breeding.

Exam question report

People with diabetes cannot lower their blood sugar levels naturally. What is least likely to affect the glucose levels of a diabetic? (1)

A a diet including high proportions of sweets and cakes
B a high protein diet
C daily injections of insulin
D daily injections of glucagon

Answer: The correct answer is B.

How students answered

Not many students got the right answer. Some thought it was C or D, but insulin and glucagon are both hormones that affect the amount of glucose in the blood, so injections of these *will* affect glucose levels.

To get it right students had to know that sweets and cakes contain carbohydrates and will be broken down to form glucose in the digestive system. Proteins are broken down to form amino acids, not glucose, so a high protein diet will have the least effect on glucose levels.

Exam question report

A dwarf variety of wheat has only recessive alleles for height. Two wheat plants, heterozygous for height, were bred together. What percentage of the offspring would be dwarf wheat plants? (1)

A 0%   **B** 25%   **C** 50%   **D** 100%
Answer: The correct answer is B.

How students answered

More than half the students got this wrong.
Students who got this right probably drew a Punnett square. You can use **H** and **h** to represent the alleles for height. The question says that the parent plants are heterozygous for height, so they must both be **Hh**. The dwarf offspring will be the ones with the **hh** genotype. The square shows that only one quarter (25%) of the offspring will have this genotype.

	H	h
H	HH	Hh
h	Hh	hh

Chemistry 1
Chemistry in our world

 In 1971, Russian geologists were drilling in the Karakum desert in Turkmenistan – an area rich in valuable minerals. They accidentally drilled into a huge rock cavern filled with natural gas. The cavern collapsed and enormous volumes of methane and poisonous gases started to escape. It was decided to set the gases alight … and they are still burning today!

In this unit you will learn about some of the natural processes that have shaped the Earth's rocks and atmosphere. You will also learn about how we obtain and use the Earth's natural resources for the manufacture of materials, and how this affects the environment and society.

Learning Outcomes

Throughout this unit you will be required to:

0.1 Recall the formulae of elements and simple compounds in the unit

0.2 Represent chemical reactions by word equations (H) and simple balanced equations

(H) *0.3* Write balanced equations including the use of state symbols (s), (l), (g) and (aq) for a wide range of reactions in this unit

0.4 Assess practical work for risks and suggest suitable precautions for a range of practical scenarios for reactions in this unit

0.5 Demonstrate an understanding that hazard symbols used on containers:

    a   indicate the dangers associated with the contents

    b   inform people about safe-working procedures with these substances in the laboratory

What was the Earth's early atmosphere like?

Figure A was taken by a space probe on the surface of Titan, one of Saturn's moons. Scientists are interested in Titan's atmosphere because they think it may be very similar to Earth's early atmosphere.

A *The surface of Titan*

1 a Name the gases thought to have been in the Earth's early atmosphere.
b What evidence supports this?

2 Some scientists believe that the Earth's early atmosphere was like that of Mars and Venus today. Explain why.

3 Some scientists believe that the Earth's early atmosphere was like that of Titan today. How is Titan's atmosphere different from that of Mars and Venus today?

The young Earth

By studying the Earth and other planets and moons, scientists hope to discover more about the Earth's early **atmosphere**.

As we shall see in C1.2, scientists think that the evolution of life on Earth caused its atmosphere to change. Scientists study planets and moons on which they think there is no life because these atmospheres have probably not changed for billions of years and therefore may resemble Earth's early atmosphere.

Scientists are particularly interested in volcanoes because they release large quantities of gases. Volcanoes on Earth today release mainly carbon dioxide and water vapour, along with small amounts of ammonia, methane and nitrogen. Many scientists believe that these gases would have been in our early atmosphere too. However, they cannot be certain how much of these gases early volcanoes produced.

The atmosphere of Titan is 98% nitrogen, which some scientists think was released by volcanoes. They think that the Earth's early atmosphere may have been like this.

There are also volcanoes on Mars and Venus but their atmospheres are mainly carbon dioxide. This has led other scientists to think that Earth's early atmosphere was mainly carbon dioxide.

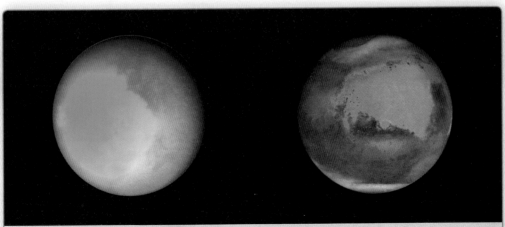

B *Was the Earth's original atmosphere more like that of Titan (left) or Mars (right)?*

Recent results from space probes have shown that Titan has an icy interior rather than a rocky one (like Earth, Mars and Venus). This makes it more likely that the Earth's early atmosphere resembled that of Mars or Venus. However, this does not explain how the Earth's atmosphere later came to contain so much nitrogen. Scientists can't be certain which theory is correct.

Oxygen

Scientists are more certain that there was little or no oxygen in the early atmosphere on Earth. There is evidence for this idea – for example, volcanoes do not release oxygen. Also the iron compounds found in the Earth's oldest rocks are compounds that would only form in the absence of oxygen.

As the Earth became older, it cooled down. The water vapour in the hot atmosphere also cooled down and is thought to have condensed to liquid water. This formed the oceans.

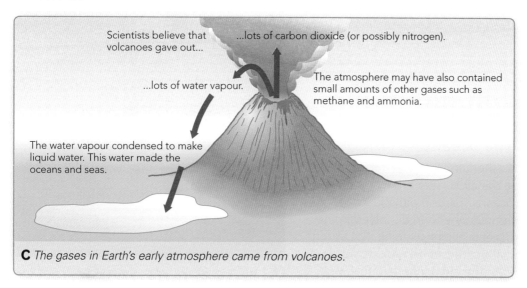

Scientists believe that volcanoes gave out...

...lots of carbon dioxide (or possibly nitrogen).

...lots of water vapour.

The atmosphere may have also contained small amounts of other gases such as methane and ammonia.

The water vapour condensed to make liquid water. This water made the oceans and seas.

C *The gases in Earth's early atmosphere came from volcanoes.*

We have only limited evidence about the Earth's atmosphere when it was young. This means that there are uncertainties about these theories on the Earth's early atmosphere.

4 Why is it difficult to be precise about the evolution of the Earth's atmosphere?

5 Why are scientists interested in Titan's atmosphere?

6 State two pieces of evidence that led scientists to think there was little or no oxygen in Earth's early atmosphere.

7 Why haven't the atmospheres on Mars or Venus changed for billions of years, but the atmosphere on Earth has?

8 Describe how the oceans formed on the Earth.

Skills spotlight

Does the evidence that scientists have found show that the early atmosphere was mainly carbon dioxide, nitrogen or both? Explain your answer.

ResultsPlus
Watch Out!

It is difficult to be certain about how the planet has changed. Be clear that you are writing about theories and express alternative points of view.

Learning Outcomes

1.1 Recall that the gases produced by volcanic activity formed the Earth's early atmosphere

1.2 Recall that the early atmosphere contained: **a** little or no oxygen **b** a large amount of carbon dioxide **c** water vapour and small amounts of other gases

1.3 Explain why there are different sources of information about the development of the atmosphere which makes it difficult to be precise about the evolution of the atmosphere

1.4 Describe how condensation of water vapour formed oceans

HSW **4** Identify questions that science cannot currently answer, and explain why these questions cannot be answered

How has the Earth's atmosphere changed?

On 21 August 1986, 1700 people died from suffocation by carbon dioxide released from Lake Nyos in Cameroon, Africa. Magma (molten rock) under the lake was releasing carbon dioxide, which dissolved in the lake water. A layer of water at the bottom of the lake got saturated with carbon dioxide, meaning there was as much dissolved as possible. More gas became trapped under this layer. In 1986, a landslide released about 1.6 million tonnes of this gas.

A *Lake Nyos is now being 'degassed' to make it safer.*

The oceans

As the Earth cooled, water vapour in the air condensed to form the oceans. Carbon dioxide in the atmosphere then dissolved in the oceans. Scientists think that about half of the carbon dioxide was lost from the atmosphere in this way. Some marine organisms – such as coral, molluscs and star fish – use dissolved carbon dioxide to make shells of calcium carbonate. As these creatures died, their shells fell and became sediment. Over millions of years, all these layers of sediment became **sedimentary rock**. **Limestone** is mostly calcium carbonate.

?

1 Limestone is a sedimentary rock. What is the main substance in limestone?

2 Much limestone is formed from the shells of marine organisms. Why do marine organisms with shells remove carbon dioxide from the atmosphere?

B *You can sometimes find shells in limestone because that's how it has formed.*

Photosynthesis

Scientists believe that life started on Earth about 4 billion years ago. About 1 billion years ago, some organisms developed the ability to **photosynthesise**, taking in carbon dioxide and releasing oxygen.

Over time, more photosynthesising organisms evolved, including plants. Increasing levels of photosynthesis sped up the rate at which carbon dioxide was removed from the atmosphere and the rate at which oxygen was added.

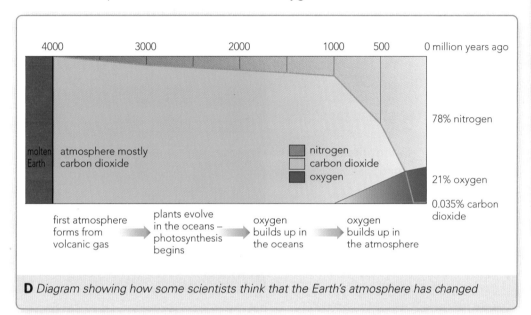

D *Diagram showing how some scientists think that the Earth's atmosphere has changed*

C *These sedimentary structures are called 'stromatolites'. Some are well over 2 billion years old and contain fossils of simple organisms that released oxygen.*

3 What are two ways that carbon dioxide was removed from the early atmosphere?

4 Explain how the amount of oxygen in the atmosphere gradually increased.

5 There are different theories about the Earth's early atmosphere (see C1.1). Look back and decide which theory the scientists who prepared Figure D support.

6 Explain how carbon dioxide from the early atmosphere ended up in sedimentary rocks.

Skills spotlight

Make a table summarising the scientific ideas about why the amounts of carbon dioxide in the atmosphere changed:
– factors that add carbon dioxide to the atmosphere
– factors that remove carbon dioxide from the atmosphere

Learning Outcomes

1.5 Describe how the amount of carbon dioxide in the atmosphere was reduced by:
  a the dissolution of carbon dioxide into the oceans
  b the later incorporation of this dissolved carbon dioxide into marine organisms which eventually formed carbonate rocks

1.6 Explain how the growth of primitive plants used carbon dioxide and released oxygen by photosynthesis and consequently the amount of oxygen in the atmosphere gradually increased

HSW 3 Describe how phenomena are explained using scientific theories and ideas

C1.3 Oxygen in the atmosphere

How much oxygen is there in the air?

Oxygen is a very important gas in air that is essential for humans. If our bodies don't have enough oxygen, it can cause headaches, sickness and eventually death. However, too much oxygen is also bad for us. Babies that are born early often have underdeveloped lungs and struggle to breathe properly. Doctors started to place premature babies in incubators where there is a lot of oxygen, but they later found that some of these babies went blind. This was due to the high levels of oxygen. Now premature babies are given very carefully controlled levels of oxygen. It is important that we can measure how much oxygen there is in an atmosphere.

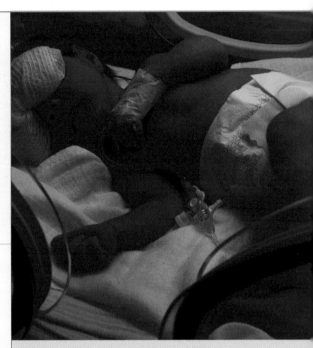

A Some premature babies have been blinded by too high a level of oxygen in incubators.

Many substances, including metals, react with the oxygen in air. Some metals react quickly and some slowly. For example, when iron rusts it uses up the oxygen in the air. Reactions such as these can be used to find the percentage of oxygen in air.

B This wrecked ship in Ireland has rusted due to the oxygen in the air.

Your task

You are going to plan an investigation that will allow you to find out how much of the air is used up when a metal reacts with the air. You are going to test the hypothesis that about one fifth of the volume of air is oxygen. Your teacher will provide you with some materials to help you organise this task.

Learning Outcomes

1.7 Investigate the proportion of oxygen in the atmosphere

ResultsPlus
Build Better Answers

When completing an investigation like this, one of the skills you will be assessed on is your ability to *choose the right equipment for your practical.* There are 4 marks available for this skill. Here are two student extracts that focus on this skill. Other skills that you need for the assessment are dealt with in other lessons.

Student extract 1 | A basic response for this skill

Use the right words for the equipment – a boiling tube is what you need here.

> For my practical I will put the iron wool into the big tube and wet it a bit. I will then turn it upside down and put it in the beaker which I will then fill with water. I will watch the water to see if it goes up the tube.

You don't need to describe the practical here – just explain what equipment you will need and why you will need it.

Student extract 2 | A good response for this skill

It is a good idea to make a list of all the equipment you think you will need.

Try to be specific about the equipment you need – in this case a stand, boss and clamp would hold the boiling tube in place.

> To carry out my investigation I will need beakers, boiling tubes, glass rod, iron/steel wool, ruler, marker pen, something to hold the boiling tube. The beaker will be used to hold the water. I will use the boiling tube to hold the iron wool. The iron wool is used because it reacts with the air to form rust (which is an iron oxide). Iron wool is good because it has a large surface area and is safe to use. I will need a marker pen because I will mark the boiling tube with the level of the water at the start of the practical and then again after a few days. By measuring the gap with the ruler I will know how much air has been used up from the boiling tube.

You need to explain why you are using each piece of equipment.

ResultsPlus

To access 2 marks
- Identify some relevant pieces of equipment
- Describe the reasons for your choices

To access 4 marks
- Identify most of the equipment that is required for your method
- Describe the reasons for all your equipment choices

What is the atmosphere like now?

Oxygen bars are very popular in Japan. They are often found in cities with polluted air. Customers breathe in air with a high concentration of oxygen for a few minutes. People find the experience relaxing and refreshing.

A *Relaxing in an oxygen bar*

The composition of the atmosphere

Nitrogen and oxygen are the most abundant gases in the atmosphere, making up about 99% between them. Small amounts of other gases make up the other 1%. There is also water vapour present in the atmosphere. The amount of water vapour changes from day to day, so it is not included when giving the composition of dry air.

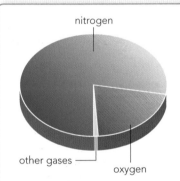

Gas	Formula	% in dry air
nitrogen	N_2	78
oxygen	O_2	21
argon	Ar	0.9
carbon dioxide	CO_2	0.04
other gases		traces

B *The gases in dry air*

Some of the other 1% of gases in the atmosphere are unreactive **noble gases**, mainly argon. A small amount of carbon dioxide is also found. There are also **trace** amounts of other gases in the atmosphere including carbon monoxide, methane, nitrogen oxides and sulfur dioxide.

The amounts of these gases in the atmosphere can vary. Some of the changes in the quantities of these gases have natural causes. For example, volcanoes can release a lot of sulfur dioxide and lightning can produce nitrogen oxides.

Other changes in the amounts of these gases are caused by human activities. **Deforestation** means that there are fewer trees to remove carbon dioxide from the atmosphere by photosynthesis. Burning fossil fuels increases the amounts of carbon dioxide, carbon monoxide and sulfur dioxide in the atmosphere. Engines and furnaces can release nitrogen oxides. Cattle and rice fields release large quantities of methane.

Some of these gases are harmful to people and/or the environment. It is important that scientists monitor how much of these gases are in air.

1 a List the three main gases in air, in order of abundance. Put the most abundant gas first.
b Draw a bar chart to show the three most abundant gases in air.

This jungle is being cleared for timber and to make farmland. Burning the jungle produces carbon dioxide and also means that there are fewer trees to remove carbon dioxide from the atmosphere.

The cattle that live on the cleared land release a lot of methane as they digest their food.

C *Deforestation is changing the atmosphere.*

Formation of nitrogen

Nitrogen is the main gas in the atmosphere today. It makes up 78% of the air. There are different theories about where it has come from.

One theory is that volcanoes released nitrogen when the Earth was young – this means that the atmosphere has always contained a lot of nitrogen. Another theory is that most of the nitrogen was added to the atmosphere gradually due to the reactions of nitrogen-containing compounds released from volcanoes.

2 Give one example of a natural activity that could increase the amount of each of these gases in the atmosphere:
a sulfur dioxide b nitrogen oxides c methane.

3 a Give two examples of human activities that could increase the amount of carbon dioxide in the atmosphere.
b Many of the trees removed by deforestation are burned. Give two reasons why deforestation increases the amount of carbon dioxide in the atmosphere.

4 Small, local changes happen in the composition of the Earth's atmosphere. Describe how some of these changes are caused by natural and human activities.

Skills spotlight

The concentration of carbon dioxide in the atmosphere can change because of natural processes and also through human activity. Suggest how scientists could measure how global levels of carbon dioxide are changing. Then explain why it is important that scientists from different places share their results.

Results**Plus**
Watch Out!

Nitrogen makes up approximately $\frac{4}{5}$ of the atmosphere and oxygen approximately $\frac{1}{5}$. All other gases, including carbon dioxide, are the remaining 1%.

Maths skills

Figure B is a **pie chart**.

Pie charts are used to show how the total is split up between different categories. The area of each sector represents the size of each category.

When drawing a pie chart you have to work out the sector angle for each category:

$$\text{sector angle} = \frac{\text{frequency} \times 360°}{\text{total frequency}}$$

E.g. Oxygen sector angle in

Figure B = $\frac{21 \times 360°}{100}$ = 75.6°

Learning Outcomes

1.8 Describe the current composition of the atmosphere and interpret data sources showing this information

1.9 Demonstrate an understanding of how small changes in the atmosphere occur through: a volcanic activity b human activity, including the burning of fossil fuels, farming and deforestation

HSW 14 Describe how scientists share data and discuss new ideas, and how over time this process helps to reduce uncertainties and revise scientific theories

How are different rocks formed?

In 1978, scientists working in Tanzania excavated a trail of fossilised footprints from our ancestors – creatures that lived 3.6 million years ago. They had walked through a layer of ash from a volcano. This was later covered by more layers of ash, preserving the footprints. The layers gradually became harder and eventually they turned into rock. More recently, erosion revealed some of the prints.

A *Fossilised footprints of two adults and a child*

B *Granite has large crystals of different colours.*

Igneous rocks

Rocks deep inside the Earth may become hot enough to partially melt. Molten rock is called **magma**. It may stay inside the Earth, or it may erupt onto the surface as **lava**. When molten rock cools down it **solidifies** and becomes solid rock. Rocks formed this way are called **igneous rocks**. They contain crystals that interlock.

The size of the **crystals** depends on the **rate** at which the magma or lava cools. The slower it cools, the larger the crystals. For example, **granite** forms from magma that cools slowly deep below the surface of the Earth. It has large crystals, whereas igneous rocks that cool quickly have small crystals.

Sedimentary rocks

1 Explain how you know that granite forms from magma that cools slowly. ②

Rocks are broken into smaller pieces by physical processes, such as the expansion of water when it freezes. They are also broken up by chemical reactions with water or air. **Erosion** happens when these pieces of rock are transported – for example in a river. Rivers carry large amounts of broken rock towards the sea, where it may become sedimentary rock.

Skills spotlight

Apply the scientific ideas on this spread to explain the difference in crystal size between granite and rhyolite. Suggest where each rock may have formed.

C *Rhyolite has a similar chemical composition to granite. Its crystals are too small to see with the naked eye.*

Most sedimentary rocks form from pieces of other rocks. Others form from the hard parts of dead organisms. **Chalk** and limestone, which are mostly calcium carbonate, can be formed from sea shells. Layers of this **sediment** build up on the sea bed. Over a long time, these layers are compacted, or squashed together, to form rock. Sedimentary rocks may contain the **fossil** remains of dead plants and animals, or imprints such as footmarks.

Metamorphic rocks

The action of heat and/or pressure can change rocks, causing new crystals to form. The changed rocks are called **metamorphic rocks**. For example, **marble** is a metamorphic rock formed from chalk or limestone. The grains in chalk and limestone are weakly joined together with small gaps between them. When marble forms, these grains become new crystals of calcium carbonate that interlock tightly. This makes marble harder than chalk or limestone.

D *Marble is a metamorphic rock formed by the action of heat and pressure.*

Erosion

Sedimentary rocks are more susceptible to erosion than igneous rocks or metamorphic rocks. This is because metamorphic rocks and igneous rocks have interlocking crystals. They are harder and less easily eroded than sedimentary rocks.

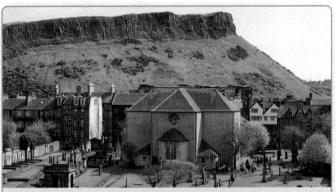

F *Salisbury Crags in Scotland are made of igneous rocks exposed by the erosion of other rock.*

E *Mylonite is a metamorphic rock formed mainly by the action of pressure.*

2 Explain what type of rock is most likely to have been eroded to produce the crags in Figure F.

3 **a** Name the main compound in chalk and limestone.
b Explain why the same compound is found in marble.

4 Name one example of each of the following types of rock and describe each type: **a** igneous rock **b** metamorphic rock **c** sedimentary rock.

5 Suggest why fossils are often found in sedimentary rocks but not in igneous rocks.

6 Describe how an igneous rock may become a sedimentary rock, which in turn may become a metamorphic rock. Explain what must happen for the metamorphic rock to become an igneous rock.

ResultsPlus
Watch Out!

To help remember that crystals are larger in rocks that cool down slowly, think of the crystals growing while the rock is hot. If the rock cools down slowly, the crystals have more time to grow bigger.

Learning Outcomes

2.1 Describe that igneous rocks, such as granite, are: **a** formed by the solidification of magma or lava **b** made of crystals whose size depends on the rate of cooling

2.2 Describe chalk and limestone as examples of sedimentary rocks

2.3 Describe how sedimentary rocks are formed by the compaction of layers of sediment over a very long time period

2.4 Recall that sedimentary rocks: **a** may contain fossils **b** are susceptible to erosion

2.5 Describe marble as an example of a metamorphic rock

2.6 Describe the formation of metamorphic rocks by the action of heat and/or pressure, including the formation of marble from chalk or limestone

2.7 Recall that limestone, chalk and marble exist in the Earth's crust and that they are all natural forms of calcium carbonate

HSW 3 Describe how phenomena are explained using scientific theories and ideas

What do cement, concrete and glass have in common?

Around 76 million tonnes of limestone are quarried in the UK every year. This is more than one tonne of the rock for each person in the UK. What uses are there for all this limestone?

A *A limestone quarry in Scotland*

Using limestone

Limestone is cut into blocks to be used for constructing buildings. It is also crushed into smaller lumps to be used to make a firm base for railway lines and roads. Limestone is a raw material for the manufacture of cement, concrete and glass. Figure B shows how much is used for each purpose.

> **1** What is the best way to present the information in Figure B and why?
>
> **2** What is the biggest use of limestone?

Use	Percentage (%)
Construction	57
Making concrete and cement	22
Other uses, e.g. making glass	16
Making steel and iron alloys	5

B *The main uses of limestone*

Quarrying limestone

Limestone is removed from the ground at a **quarry**. Explosives are used to break the limestone into pieces. These are cut or crushed into useful sizes, then taken by road or rail to the customers.

Limestone is an important material and there is a commercial need for it but quarries have benefits and drawbacks. They are in the countryside, where jobs may be difficult to find. Having jobs at a quarry helps local families and businesses. Limestone is valuable and is exported to other countries, helping the UK's economy.

Unfortunately, quarries are commonly in attractive places like the Peak District. They are dusty and noisy. Quarries may affect the quality of life for local people and damage the tourist industry. Heavy lorries cause extra traffic, noise and pollution. Land taken up by a quarry cannot be used for farming or other purposes. Quarries destroy the original landscape. However, the quarry owners usually restore it later, often as farmland or a nature reserve.

ResultsPlus
Watch Out!

If you are asked to discuss the development of a new quarry, be careful to consider both the advantages and disadvantages. Don't just offer one side of the argument.

Thermal decomposition

Limestone, chalk and marble are natural sources of calcium carbonate. When it is heated strongly, calcium carbonate breaks down (decomposes) to form calcium oxide and carbon dioxide. **Word equations** show what happens in chemical reactions. This is the word equation for calcium carbonate decomposing:

calcium carbonate → calcium oxide + carbon dioxide

This type of reaction is called **thermal decomposition**. Thermal decomposition of limestone is used in the manufacture of cement and glass.

Making cement and concrete

Cement is made by heating limestone with powdered clay. Cement is an ingredient in mortar – the mixture used by bricklayers to hold bricks together in walls. **Concrete** is made by mixing cement with sand, gravel and water. Concrete is widely used in the construction of buildings and bridges.

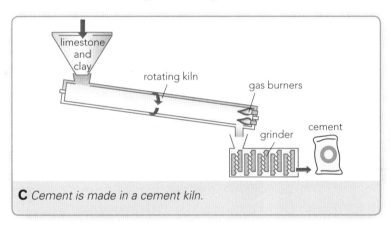

C *Cement is made in a cement kiln.*

Glass is made by heating limestone with sand and sodium carbonate. A chemical reaction occurs and liquid glass is made. This cools to form the hard, transparent solid we use for making windows.

D *Glass bottles being made from molten glass*

3 Write a word equation for the thermal decomposition of calcium carbonate.

4 Describe how each of the following is made:
a cement b concrete c glass.

5 Draw a table or spider diagram to show the advantages and disadvantages of quarrying limestone.

6 Explain why an increase in demand for new housing may cause an increase in demand for limestone.

Learning Outcomes

2.8 Demonstrate an understanding of the balance between the demand for limestone and the economic, environmental and social effects of quarrying it

2.9 Demonstrate an understanding of the commercial need for quarrying calcium carbonate on a large scale, as a raw material for the formation of glass, cement and concrete

2.10 Describe the thermal decomposition of calcium carbonate into calcium oxide and carbon dioxide

HSW **10** Use qualitative and quantitative approaches when presenting scientific ideas and arguments, and recording observations

C1.7 Thermal decomposition of carbonates

Do all metal carbonates break down in the same way?

Two thousand years ago, the Romans knew that calcium carbonate forms a useful product when it is heated and were using this knowledge. Archaeologists have found Roman 'lime kilns' all over Europe. These kilns were used to heat limestone to make calcium oxide, which was used to make mortar and concrete.

A *The Pantheon in Rome was built almost two thousand years ago. Its 43.3 m dome is still the largest non-reinforced concrete dome in the world.*

B *Carbon dioxide turns limewater milky.*

Calcium carbonate decomposes when it is heated, producing calcium oxide and carbon dioxide. Carbon dioxide can be detected using limewater. When the carbon dioxide is passed through it the limewater turns milky.

Your task

You are going to plan an investigation that will allow you to find out how easily different metal carbonates decompose. You are going to test the hypothesis that some metal carbonates decompose more easily than others when they are heated. Your teacher will provide you with some materials to help you organise this task.

Learning Outcomes

2.11 Investigate the ease of thermal decomposition of carbonates, including calcium carbonate, zinc carbonate and copper carbonate

When completing an investigation like this, one of the skills you will be assessed on is your ability to *collect and record primary data.* There are 4 marks available for this skill. Here are three student extracts that focus on this skill. Other skills that you need for the assessment are dealt with in other lessons.

Student extract 1 — A basic response for this skill

There are no units here.

> Copper carbonate – 15, zinc carbonate – 54

Here we can see that the student has collected some results but that the full range has not been tested.

Student extract 2 — A better response for this skill

A simple table or a list helps to make your results easier to read.

> Copper carbonate 15 s
> Zinc carbonate 54 s – some of the solid moved up the tube during heating
> Calcium carbonate – the limewater did not go milky by 2 minutes

There is a better range of results here.

Student extract 3 — A good response for this skill

A better way of arranging results is in a table.

| Name of metal carbonate | Time taken for limewater to turn milky (s) | | Observations |
	Experiment 1	Experiment 2	
Copper carbonate	15	19	Some powder moved up the tube in the second experiment
Zinc carbonate	54	48	Some powder moved up the tube in the first experiment
Calcium carbonate	Did not turn milky	Did not turn milky	

The table needs clear headings and units.

Repeat results mean even more evidence to use when discussing the hypothesis.

ResultsPlus

To access 2 marks
- Collect a suitable range of data or observations
- Record some of these appropriately

To access 4 marks
- Collect a suitable range of data or observations
- Record all your evidence appropriately
- Record repeat data or further relevant data

::: What happens to the atoms in chemical reactions?

Zinc carbonate is used as a fireproof ingredient in some plastics. In a fire, the zinc carbonate absorbs some of the heat. As it decomposes, it releases carbon dioxide. This helps to reduce the amount of oxygen available near the surface of the burning plastic. The zinc oxide left behind helps to form a layer of heat-resistant material. Zinc carbonate is chosen instead of calcium carbonate because it decomposes more easily when it is heated.

A *Fireproof materials help to protect firefighters.*

Word equations

Word equations show what happens in chemical reactions. In general:

reactants → products

In the thermal decomposition of zinc carbonate, the **reactant** is zinc carbonate. The **products** are zinc oxide and carbon dioxide. Here is the word equation:

zinc carbonate → zinc oxide + carbon dioxide

Atoms and chemical reactions

Substances are made of **atoms**. An atom is the smallest part of an **element** that can take part in chemical reactions. A **compound** consists of the atoms of two or more different elements chemically joined together. The **chemical formula** of a compound shows the symbols of the elements it contains and the ratios in which their atoms are present.

In all chemical reactions, the atoms of the reactants rearrange to form new products. None of the atoms are destroyed in the reaction and no new ones are formed. The products have different physical and chemical properties from those of the reactants because their atoms are combined differently.

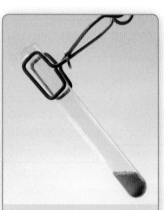

B *Green copper carbonate decomposes very easily when it is heated – it forms black copper oxide and carbon dioxide.*

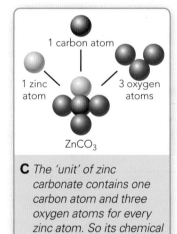

1 carbon atom

1 zinc atom

3 oxygen atoms

$ZnCO_3$

C *The 'unit' of zinc carbonate contains one carbon atom and three oxygen atoms for every zinc atom. So its chemical formula is $ZnCO_3$.*

(?)

1 Write the word equation for the thermal decomposition of copper carbonate.

2 What is an atom?

3 The formula for calcium carbonate is $CaCO_3$.
a How many different elements does calcium carbonate contain?
b What does the '3' in the formula tell you?

Balanced equations

Balanced equations show what happens to the atoms in a chemical reaction. Here is the balanced equation for the thermal decomposition of zinc carbonate:

zinc carbonate → zinc oxide + carbon dioxide
$ZnCO_3(s)$ → $ZnO(s)$ + $CO_2(g)$

There is one Zn, one C and three O on each side of the equation. Each of these symbols represents an atom of an element. There are the same numbers of atoms of each element on both sides, so it is a *balanced* equation.

H

The symbols (s) and (g) are **state symbols**. They show that $ZnCO_3$ and ZnO are solids and CO_2 is a gas. Pure liquids are shown by (l). **Aqueous solutions**, formed when substances dissolve in water, are shown by (aq).

(H) **4** Write a balanced equation for the thermal decomposition of calcium carbonate to calcium oxide, CaO, and carbon dioxide, CO_2.

Conserving mass

Atoms are not made or destroyed in a chemical reaction – they are only rearranged. So the total mass before and after a reaction is the same. This works for all reactions. It is easiest to demonstrate in reactions where none of the reactants or products are allowed to leave the system. This happens if the reaction takes place in a stoppered test tube or in a solution where none of the substances leaves the liquid.

Precipitation reactions

Precipitation reactions happen when **soluble** substances react together to form an **insoluble** product, called the **precipitate**. For example, silver nitrate and potassium bromide are soluble. Their solutions react together to form insoluble cream-coloured silver bromide:

silver nitrate + potassium bromide → potassium nitrate + silver bromide

(H) $AgNO_3(aq)$ + $KBr(aq)$ → $KNO_3(aq)$ + $AgBr(s)$

Skills spotlight

(H) Describe the advantages and disadvantages of using balanced equations rather than word equations.

 **ResultsPlus**
Watch Out!

(H) NaCl(aq) means sodium chloride disolved in water (aqueous solution).

5 A student mixes 2 g of a lead nitrate solution with 2 g of a potassium iodide solution. A precipitation reaction occurs. What is the total mass of the mixture after the reaction?

6 When 10.0 g of calcium carbonate is heated strongly, it decomposes completely to form 5.6 g of calcium oxide. What mass of carbon dioxide must have been released?

7 The table shows temperatures at which different carbonates decompose:

Carbonate	Temperature (°C)
calcium carbonate	825
copper carbonate	200
sodium carbonate	1000
zinc carbonate	300

a Use this information to put the carbonates in order of ease of decomposition – start with the one that is most difficult to decompose.

b Some fireproofing materials produce carbon dioxide, which can help to put out fires if they start. From the table, choose the most suitable substance for this purpose and explain how it works.

Learning Outcomes

2.12 Describe the ease of thermal decomposition of different metal carbonates

2.13 Demonstrate an understanding that: **a** atoms are the smallest particles of an element that can take part in chemical reactions **b** during chemical reactions, atoms are neither created nor destroyed **c** during chemical reactions, atoms are rearranged to make new products with different properties from the reactants

2.16 Demonstrate an understanding that the total mass before and after a reaction in a sealed container is unchanged, as shown practically by a precipitation reaction

HSW **11** Present information using scientific conventions and symbols

How does calcium carbonate react with other substances?

Concrete releases heat as it sets. The Hoover Dam was built in the USA in 1936 and so much concrete was used that it would have taken 125 years to cool down! To speed this up, cooling water was passed through pipes set in the surrounding concrete to help to cool the setting concrete.

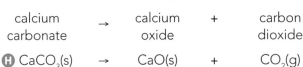

A *The Hoover Dam contains over $3\,300\,000\,m^3$ of concrete.*

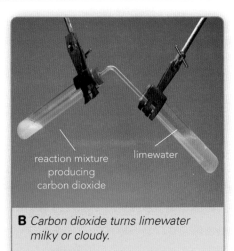

reaction mixture producing carbon dioxide

limewater

B *Carbon dioxide turns limewater milky or cloudy.*

Making limewater

Limestone is a raw material used in the manufacture of concrete. When limestone is heated, the calcium carbonate that it contains forms calcium oxide.

| calcium carbonate | → | calcium oxide | + | carbon dioxide |

 $CaCO_3(s)$ → $CaO(s)$ + $CO_2(g)$

A vigorous reaction happens when water is added to calcium oxide. A lot of heat is released, which makes the water boil as it touches the calcium oxide. Calcium hydroxide, a crumbly white solid, forms in the reaction:

| calcium oxide | + | water | → | calcium hydroxide |

Ⓗ $CaO(s)$ + $H_2O(l)$ → $Ca(OH)_2(s)$

Calcium hydroxide dissolves when more water is added, forming calcium hydroxide solution. This solution is often called **limewater**.

Ⓗ The brackets in the formula $Ca(OH)_2$ show that each 'unit' of calcium hydroxide contains two hydroxide 'units'. So a 'unit' of calcium hydroxide contains one calcium atom, two oxygen atoms and two hydrogen atoms.

A test for carbon dioxide

Limewater turns cloudy (milky) in the presence of carbon dioxide. This is because white insoluble calcium carbonate forms:

| calcium hydroxide | + | carbon dioxide | → | calcium carbonate | + | water |

Ⓗ $Ca(OH)_2(aq)$ + $CO_2(g)$ → $CaCO_3(s)$ + $H_2O(l)$

ResultsPlus
Watch Out!

Atoms are not used up in chemical reactions. They are rearranged to make new products instead.

1 Name the precipitate made when carbon dioxide reacts with limewater.

If a lot of carbon dioxide is bubbled through the limewater, the calcium carbonate disappears and a colourless solution is formed. This happens because carbon dioxide dissolves in water to form an acidic solution, which reacts with the calcium carbonate. Other acids react with calcium carbonate, which is useful for farmers and coal-fired power stations.

Neutralising acids with limestone

Acids are neutralised by alkalis. This is called a **neutralisation reaction**. Calcium carbonate, calcium oxide and calcium hydroxide can neutralise acids. Some crops do not grow well if the soil is too acidic. Farmers may need to reduce the acidity of their soil. They spray powdered calcium carbonate, calcium oxide or calcium hydroxide over their fields to do this.

Many power stations burn coal. Coal naturally contains sulfur and sulfur compounds. When the coal burns, the sulfur forms sulfur dioxide:

sulfur	+	oxygen	→	sulfur dioxide
ⓗ $S(s)$	+	$O_2(g)$	→	$SO_2(g)$

Nitrogen oxides are also formed when the coal burns. Sulfur dioxide and nitrogen oxides are acidic gases. They produce acid rain if they escape from the chimneys into the atmosphere. To stop this happening, wet powdered calcium carbonate is sprayed through the waste gases. This reacts with the acidic gases and neutralises them. In this way, limestone reduces harmful emissions and helps to reduce acid rain.

2 Describe what happens when water is added to calcium oxide.

3 a State the common name for the solution formed when calcium hydroxide dissolves in water. b Describe a laboratory test for carbon dioxide using this solution.

4 Calcium carbonate breaks down to form calcium oxide and carbon dioxide when heated. Calcium oxide reacts with water to form calcium hydroxide. This reacts with carbon dioxide to form calcium carbonate again, and water. Write equations to describe these three reactions.

5 Which calcium compounds can be used to reduce soil acidity? a Explain how they reduce the acidity. b Explain why farmers need to control soil acidity.

6 Describe the use of calcium carbonate in reducing harmful emissions from coal-fired power stations.

Learning Outcomes

2.14 Describe the effect of water on calcium oxide

2.15 Describe how calcium hydroxide dissolves in water to form a solution, known as limewater

2.17 Explain how calcium oxide, calcium hydroxide and calcium carbonate can be used to neutralise soil acidity

2.18 Explain how calcium carbonate can be used to remove acidic gases from coal-fired power station chimneys, reducing harmful emissions and helping to reduce acid rain

HSW 5 Plan to test a scientific idea, answer a scientific question, or solve a scientific problem by selecting appropriate data to test a hypothesis

>>>>>>>>>>>>>>>>>>>>>>> Why would eating chalk be helpful for an upset stomach?

How do indigestion remedies work?

Traditional Chinese medicine uses powdered pearls to cure many things, including indigestion. Pearls are mostly made from calcium carbonate and this substance helps to neutralise the extra acid that causes indigestion.

A *Pearls grow inside oysters.*

1 Write down two reasons why your stomach produces acid.

2 What causes indigestion?

3 What is an antacid?

Looking after your stomach

Your stomach produces hydrochloric acid (HCl), which kills bacteria that may be on your food. Food in the stomach is **digested** (broken down) by digestive enzymes that need acidic conditions to work properly.

Sometimes the stomach produces too much acid and this can cause stomach pain we call indigestion. Sometimes the acid can escape from the top of the stomach, which causes pain in the tube leading to the mouth. This is known as heartburn, although it has nothing to do with the heart.

Medicines called **antacids** can neutralise excess stomach acid. Antacids contain **bases** – substances that can react with acids. Figure C shows some common bases that are used in antacids.

ResultsPlus
Watch Out!

Stomach acid will *always* be hydrochloric acid. When it is neutralised by a base, salt and water are produced.

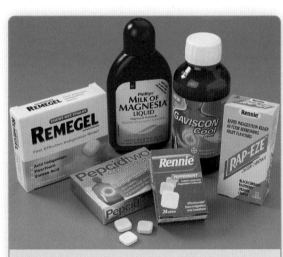

B *Different indigestion remedies*

The neutralisation reaction between an acid and a base produces water and a compound called a **salt**. This can be shown by the general equation:

acid + base → salt + water

Some bases are soluble. A base dissolved in water is called an **alkali**.

Compound	Formula
calcium carbonate	$CaCO_3$
magnesium carbonate	$MgCO_3$
sodium hydrogencarbonate (sodium bicarbonate)	$NaHCO_3$
magnesium hydroxide	$Mg(OH)_2$
aluminium hydroxide	$Al(OH)_3$

C *Some antacid compounds*

If an indigestion tablet contains magnesium hydroxide, the neutralisation reaction is:

hydrochloric acid + magnesium hydroxide → magnesium chloride + water

H $2HCl(aq)$ + $Mg(OH)_2(s)$ → $MgCl_2(aq)$ + $2H_2O(l)$

However, if the antacid contains a carbonate, carbon dioxide is also produced:

acid + carbonate → salt + water + carbon dioxide

So for a tablet containing calcium carbonate, the neutralisation reaction will be:

hydrochloric acid + calcium carbonate → calcium chloride + water + carbon dioxide

H $2HCl(aq)$ + $CaCO_3(s)$ → $CaCl_2(aq)$ + $H_2O(l)$ + $CO_2(g)$

We can describe acids and alkalis using the **pH scale**. A neutral liquid (such as water) has a pH of 7. The range of pH values of acids and alkalis is shown in Figure D.

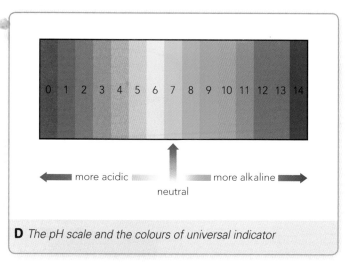

more acidic ← ← neutral → → more alkaline

D *The pH scale and the colours of universal indicator*

We can find out if a liquid is an acid or an alkali using an **indicator**. Figure D shows the colours of **universal indicator** at each point on the pH scale. **Litmus paper** is an indicator. Blue litmus paper turns red under acid conditions and red litmus paper turns blue under alkaline conditions. Litmus paper must be damp to work.

4 Sodium hydroxide is an alkali. Is sodium hydroxide soluble? Explain your answer.

5 Stomach acid would turn universal indicator orange. What is its pH?

6 Write word equations for the reactions between hydrochloric acid and:
a aluminium hydroxide
b magnesium carbonate.

H 7 Write balanced equations for the reactions in question 6. Some of the formulae you need are Al(OH)₃, AlCl₃, MgCO₃, MgCl₂.

8 Write a short paragraph to explain what causes indigestion and how it is cured.

Learning Outcomes

3.1 Recall that hydrochloric acid is produced in the stomach in order to:
  a help digestion
  b kill bacteria

3.2 Describe indigestion remedies as containing substances that neutralise excess stomach acid

HSW 12 Describe the benefits, drawbacks and risks of using new scientific and technological developments

Are there medicines in your kitchen cupboards?

C1.11 Indigestion remedies

How can you find out which indigestion remedies work the best?

Many people buy indigestion tablets from a chemist. However, some people think this is a waste of money and use cooking ingredients found in their kitchen cupboards.

Stomach pains are often caused by indigestion, when the stomach produces too much acid. Indigestion remedies contain substances that neutralise the excess acid in the stomach to make you feel better. However some people recommend just using a teaspoonful of baking powder or bicarbonate of soda, which you use for cooking.

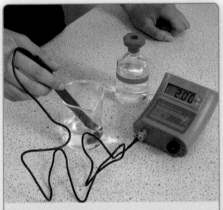

B *The acid in the beaker is similar to the acid normally in your stomach. The pH meter shows how acidic it is.*

A *The indigestion remedy on the top costs up to four tir less than the store-bought antacid on the bottom.*

Your task

You are going to plan an investigation which will allow you to find out about indigestion remedies. You are going to test the hypothesis that indigestion remedies bought from a chemist are better than using baking powder or bicarbonate of soda. Your teacher will provide you with some materials to help you organise this task.

Learning Outcomes

3.3 Investigate the effectiveness of different indigestion remedies

Build Better Answers

When completing an investigation like this, one of the skills you will be assessed on is your ability to *control variables*. There are 6 marks available for this skill. Here are two student extracts that focus on this skill. Other skills that you need for the assessment are dealt with in other lessons.

Student extract 1 — A typical response for this skill

> In this experiment I need to control the concentration of the acid, the volume of the acid, the amount the mixture is stirred and the amount of indigestion remedy that I use. I will do this by using the same concentration of acid each time, measuring the volume of acid using a measuring cylinder, stirring the mixture for the same amount of time in each practical and using one dose of each indigestion remedy (in the case of the bicarbonate of soda this will be one teaspoon because this is what is recommended in the text book.)

This is a clear list of variables and covers most of the important ones in this practical.

These are clear decriptions of how to control the variables.

Student extract 2 — A good response for this skill

> I need to control the concentration of the acid, the volume of the acid, the amount the mixture is stirred, the temperature, the surface area and the amount of indigestion remedy that I use. I will do this by using the same concentration of acid each time. If I use the same bottle of acid each time then this will make sure that I am using the same concentration of acid. If I measure the volume of acid carefully using a measuring cylinder this will mean that I am using the same amounts each time. To keep the stirring of the mixture consistent I will stir the mixture for 30 seconds as I add the acid. I will use one dose of each indigestion remedy (in the case of the bicarbonate of soda this will be one teaspoon because this is what is recommended in the text book). I will use whatever is the recommended dose on the packet each time. I will also crush the tablets up so that they are a powder because the bicarbonate of soda is a powder. By crushing the tablets it will mean that each of the remedies has a similar surface area for the acid to react with.

This is the full range of variables which are involved in this practical.

As well as describing how you keep each variable under control you also need to explain why you are going to do it this way.

ResultsPlus

To access 4 marks
- Identify some relevant variables to control
- Describe how to appropriately control these variables

To access 6 marks
- Identify a range of relevant variables to control
- Explain how to control these variables

How are acids neutralised?

Many different substances are transported around the country. The emergency services need to know how to deal with spilled substances that might be harmful if there is an accident. Any lorries or tankers carrying these substances must have a warning sign. This tells people about the dangers of the substance so they know how to clear it up.

A *This tanker spilled about 20 tonnes of methylamine, an inflammable and poisonous chemical.*

Skills spotlight

Scientists use many different internationally agreed symbols and conventions (standard ways of doing things). Why is it necessary that scientists agree on conventions?

1 If your stomach had a hazard symbol, which one would it have? Explain your answer.

2 Why are acids and many other substances stored in containers that have hazard symbols on them?

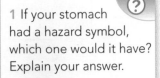

Watch Out!

Only metal *oxides*, *hydroxides* and *carbonates* can be used to neutralise acids.

Hazard symbols

A **hazardous** substance is dangerous if it is not used and stored correctly. It is important that people dealing with hazardous substances know how the substance might be dangerous. There is an international system of **hazard symbols** used on containers to show the dangers associated with the substances they contain. The labels usually have reference numbers as well, and these can be used to find more details of the hazards.

Corrosive

Caution (harmful or irritant)

Flammable

Toxic

B *Some of the internationally agreed hazard warning symbols. Bottles of acid or alkali will have the irritant or harmful or corrosive sign on them. Source: http://www.unece.org/trans/danger/publi/ghs/pictograms.html. Reproduced with the kind permission of the United Nations Economic Commisssion for Europe.*

Containers of substances used in laboratories also have hazard symbols. The symbols inform people working with the substances about the kinds of precautions that need to be taken for safe working.

Neutralising acids

If an acid or an alkali is spilled, it can be made safe by diluting it with water or by neutralising it. A neutral solution is less dangerous than either an acid or an alkali. There are three different types of compounds that can be used to neutralise acids. These are metal oxides, metal hydroxides and metal carbonates.

The general equations for these reactions are:

acid + metal oxide → salt + water
acid + metal hydroxide → salt + water
acid + metal carbonate → salt + water + carbon dioxide

The salt that is formed in a neutralisation reaction depends on the acid:
- sulfuric acid produces sulfate salts
- nitric acid produces nitrate salts
- hydrochloric acid produces chloride salts.

sulfuric acid	+	copper oxide	→	copper sulfate	+	water

Ⓗ $H_2SO_4(aq)$ + $CuO(s)$ → $CuSO_4(aq)$ + $H_2O(l)$

nitric acid	+	sodium hydroxide	→	sodium nitrate	+	water

Ⓗ $HNO_3(aq)$ + $NaOH(s)$ → $NaNO_3(aq)$ + $H_2O(l)$

hydrochloric acid	+	copper carbonate	→	copper chloride	+	water	+	carbon dioxide

Ⓗ $2HCl(aq)$ + $CuCO_3(s)$ → $CuCl_2(aq)$ + $H_2O(l)$ + $CO_2(g)$

C Clearing up an acid spill

D Copper oxide reacting with dilute sulfuric acid. The blue colour in the beaker shows that copper sulfate is being formed.

3 The workers in Figure C are making some spilled acid safe. What could the white powder be? Explain your answer.

4 Why do you see bubbles when you react copper carbonate with an acid but not when you use copper oxide?

5 Write word equations for the following reactions:
a nitric acid + copper oxide
b hydrochloric acid + calcium hydroxide
c sulfuric acid + copper carbonate

Ⓗ 6 Write balanced symbol equations for the reactions in question 5. (Some of the formulae you need are Cu(NO₃)₂, Ca(OH)₂, CaCl₂.)

7 Explain why hazard symbols are used and why it is important that everyone uses the same set of symbols.

Learning Outcomes

3.4 Recall that acids are neutralised by:
a metal oxides b metal hydroxides c metal carbonates to produce salts (no details of salt preparation techniques or ions are required)

3.5 Recall that: a hydrochloric acid produces chloride salts b nitric acid produces nitrate salts c sulfuric acid produces sulfate salts

HSW 11 Present information using scientific conventions and symbols

>>>>>>>>>>>>>>>>>>>>>>>>>>> Is neutralisation the only way to stop something being an acid?

C1.13 Electrolysis

How can an acid be broken down into its elements?

Different elements can react together to form compounds. It is usually quite difficult to make compounds split up into the original elements again. Some compounds can be decomposed (split up) using electricity. In 1807 Sir Humphry Davy discovered the elements potassium, sodium, barium, calcium and magnesium by using electricity to decompose different compounds.

A Davy (the man with the bellows) showed off his discoveries in dramatic public demonstrations at the Royal Institution, where he worked. You can still attend lectures in the same theatre at the Royal Institution today.

Hydrochloric acid is made by reacting hydrogen and chlorine together to form hydrogen chloride. This forms a gas when it dissolves in water. Substances such as hydrochloric acid can be broken down using electrical energy. This process is called electrolysis.

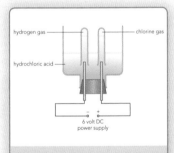

B Electrolysis can be carried out using this apparatus connected to a d.c. supply.

If you are carrying out experiments to break down compounds, it is a good idea to be able to test all the different gases that could be given off.

You can test a gas to see if it is chlorine by holding a piece of damp blue litmus paper in the mouth of the tube. If the gas is chlorine the paper will turn red and then turn white, as it is bleached. (See Figure B on C1.14). You can usually smell chlorine gas as well, but sniffing the gas given off is not a good idea as chlorine is toxic.

You can test a gas to see if it is hydrogen by holding a lighted splint in the mouth of a test tube containing the gas. If the gas is hydrogen, the mixture of hydrogen and air in the tube will explode with a squeaky 'pop'.

Your task

You are going to plan an investigation to find out how acids react when they have an electric current passed through them. You are going to test the hypothesis that all acids release hydrogen gas when they undergo electrolysis. Your teacher will provide you with some materials to help you organise this task.

Learning Outcomes

3.6 Describe electrolysis as a process in which electrical energy, from a d.c. supply, decomposes compounds, by considering the electrolysis of dilute hydrochloric acid to produce hydrogen and chlorine (explanations of the reactions at the electrodes are not required)

3.7 Investigate the electrolysis of dilute hydrochloric acid

3.8 Describe the chemical test for hydrogen

3.9 Describe the chemical test for chlorine

ResultsPlus
Build Better Answers

When completing an investigation like this, one of the skills you will be assessed on is your ability to *collect and record secondary evidence*. There are 2 marks available for this skill. Here are two student extracts that focus on this skill. Other skills that you need for the assessment are dealt with in other lessons.

Student extract 1 — A basic response for this skill

Try to find secondary evidence from at least two different sources.

Secondary evidence does not have to be data. It can be comments or observations.

> I have looked on the internet and found two different websites which are relevant to my assessed practical. The first website is from lecture notes from The University Chemistry course 100 and the notes are by a Dr AN Other. They have also been published in a scientific journal. I have noted the web address below. The second website is from a school in Britain. The pupils have done projects on electrolysis and one of the projects says 'We passed electricity through two different acids and nothing happened.' There is no name on the project.

You should be able to write down exactly where the information came from so make sure that you keep a record.

Student extract 2 — A good response for this skill

> I have looked on the internet and found two different websites which are relevant to my assessed practical. The first website is from lecture notes from The University Chemistry course 100 and the notes are by a Dr AN Other. They have also be published in a scientific journal. I have noted the web address below. The second website is from a school in Britain. The pupils have done projects on electrolysis and one of the projects says 'We passed electricity through two different acids and nothing happened.' There is no name on the project. I think that the secondary data from The University website is credible. The data in the table comes from a named scientist and is published in a scientific journal. The comment from other pupils is less credible I think. I don't know how they carried out the practical or which acids they tested.

Scientists publish data in journals and it is checked by other scientists before it is published.

If something is published on the Internet it does not mean that it is true!

ResultsPlus

To access 1 mark
- Collect and record secondary evidence which is relevant to the hypothesis in a way appropriate for the task

To access 2 marks
You also need to comment on the quality of the sources of your secondary evidence.

A *A painting by John Singer Sargent showing injured soldier after a gas attack*

What uses are there for chlorine?

Chlorine gas was used as a weapon in the First World War (1914–18). It dissolves in the moisture in the eyes and lungs to form an acid. Gassed soldiers usually died painfully within a couple of days. Figure A shows wounded soldiers after a gas attack.

In some reactions, compounds are **decomposed** (broken up) to form new compounds. Some compounds can be decomposed using heat. This is called thermal decomposition. Electricity can be used to decompose some compounds. This process is called **electrolysis**. Compounds that can be decomposed by electrolysis are **electrolytes**.

Sea water contains many dissolved substances. The most common of these is sodium chloride (common salt). If a direct current is passed through sea water, chlorine gas is produced at one of the electrodes.

You can test a gas to see if it is chlorine by holding a piece of damp blue litmus paper in the mouth of the tube. If the gas is chlorine the paper will turn red and then turn white as it is bleached.

Any acidic substance will turn damp blue litmus paper red. The key point about chlorine is that it also bleaches the litmus.

1 What is the symbol for an atom of chlorine?

2 a What is the name of the main salt dissolved in sea water?
b What is produced by the electrolysis of sea water?

3 Look at the test for chlorine. Why does the litmus paper
a turn red
b and then turn white?

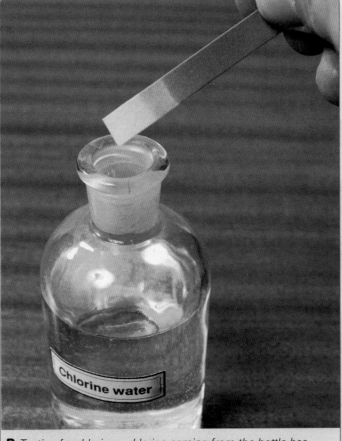
B *Testing for chlorine – chlorine coming from the bottle has turned the litmus paper red and then bleached it.*

Chlorine water

Uses of chlorine

Chlorine is a yellow-green toxic gas with many uses. You may recognise the smell of chlorine because compounds that release chlorine are used to keep water free from bacteria in swimming pools.

Chlorine is used:
- to treat our water supply, killing microorganisms
- in manufacturing bleach and other cleaning products
- in the manufacture of plastics such as poly(chloroethene), PVC.

Bleach is harmful to living things – but it also reacts with many coloured substances, converting them to colourless products. Chlorine bleaches are used in paper-making.

Skills spotlight

What are some of the benefits and drawbacks of manufacturing chlorine on a large scale? Do the benefits outweigh the risks?

Potential problems

Around 45 million tonnes of chlorine are produced each year by electrolysis of sea water. Sometimes gases can leak accidentally from a chemical factory, or a tanker carrying the gas may be involved in an accident.

Chlorine is a toxic gas and if a leak occurred near a town, many people could be killed or badly injured. A leak of toxic gas could kill more people than an explosion.

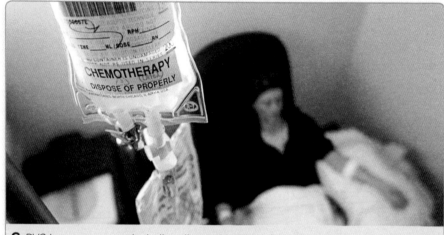

C *PVC has many uses, including disposable items for use by doctors.*

4 Write down:
a two properties of chlorine
b two uses of chlorine compounds.

5 Which property of chlorine makes it useful as a disinfectant?

6 Describe the hazard symbol that should be used on chlorine containers.

7 Write some instructions for a swimming pool attendant explaining why chlorine compounds are added to the swimming pool, why containers of the substances added have hazard signs and what some of the common hazard signs mean.

ResultsPlus
Watch Out!

Remember: the symbol for the *element* chlorine is Cl, the symbol for the *atom* chlorine is Cl, but the formula for chlorine gas is Cl_2. It is never written as Cl^2.

Learning Outcomes

3.10 Recall that chlorine can be obtained from sea water by electrolysis (explanations of the reactions at the electrodes are not required)

3.11 Describe chlorine as a toxic gas and that this leads to potential hazards associated with its large-scale manufacture

3.12 Describe the use of chlorine in the manufacture of bleach and of the polymer poly(chloroethene) (PVC)

HSW 12 Describe the benefits, drawbacks and risks of using new scientific and technological developments

What are the products of the electrolysis of water?

Nuclear submarines can stay underwater for months without coming to the surface at all. Electrolysis is used to produce oxygen for the crew to breathe.

A *This nuclear submarine can stay underwater for months.*

The electrolysis of dilute hydrochloric acid produces hydrogen gas and chlorine gas. Chlorine can also be produced by the electrolysis of sea water.

If a direct electric current is passed through water, hydrogen and oxygen are given off at the electrodes.

ResultsPlus
Watch Out!

When a small amount of acid is added to make the electrolysis of water work at a lower voltage, it is still the water that is being decomposed, not the acid.

> **1** What are the products of the electrolysis of:
> **a** dilute hydrochloric acid **b** water?
>
> **2** Write word equations to show what happens when these compounds are decomposed:
> **a** hydrogen chloride **b** water containing dilute sulfuric acid.
>
> **H** **3** Write balanced equations for the reactions in question 2.

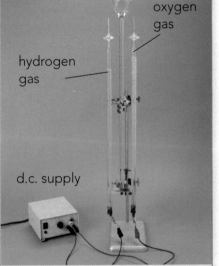

B *The electrolysis of water. The apparatus contains water with a little acid added to make the electrolysis work at low voltages.*

oxygen gas

hydrogen gas

d.c. supply

The gases are collected in test tubes. You can test the gases given off to find out what they are.

- To test for hydrogen, hold a lighted splint in the mouth of the test tube. If the test tube contains some hydrogen mixed with air it will explode with a squeaky 'pop'.
- To test for oxygen, light a splint and then blow the flame out so that the end of the splint is just glowing. Put the glowing splint in the mouth of the test tube. If the gas is oxygen, the glowing splint will relight, bursting into flame again.

Uses of hydrogen and oxygen

Hydrogen has many uses. One of the best known is as a rocket fuel. Most of the hydrogen used is made from substances obtained from oil (you will learn more about this later in this unit) but this hydrogen often contains traces of other substances. If the hydrogen has to be very pure, it is usually made by electrolysis.

We all need to breathe oxygen from the air around us. In hospitals, patients with breathing problems are often given extra oxygen.

In enclosed spaces, such as nuclear submarines and spacecraft, oxygen must either be carried in cylinders or made using electrolysis.

C *Hydrogen is the lightest gas. It is used to fill weather balloons that take measuring instruments high into the atmosphere.*

D *Welding torches produce very hot flames, melting metals to cut through them or joining pieces of metal together. Oxygen is mixed with the fuel before it burns, to make the flame very hot.*

4 How would you test a gas to see if it is:
a hydrogen
b oxygen?

5 What hazard symbol should you find on a container of hydrogen? Explain how you worked out your answer.

6 Explain some of the safety precautions you should take if you are carrying out the electrolysis of water.

Skills spotlight

Airships are filled with helium, not with hydrogen, even though using hydrogen would provide more lift force. Helium is not as easy to produce as hydrogen, but it does not burn. What are the benefits, drawbacks and risks of using hydrogen in airships? What about helium?

Learning Outcomes

3.13 Recall that water can be decomposed by electrolysis to form hydrogen and oxygen

3.14 Describe the chemical test for oxygen

HSW *12* Describe the benefits, drawbacks and risks of using new scientific and technological developments

What is the mass of the biggest lump of gold ever found?

Where do the metals we use come from?

Napoleon III, Emperor of France, is said to have held a banquet where the most important guests were given aluminium plates and cutlery. At that time, aluminium was more expensive than gold. The price of a metal depends on how it is obtained.

A *Napoelon used aluminium plates and cutlery.*

B *The largest gold nugget ever found (27.2 kg)*

Metals found as elements

A few metals are found naturally in the Earth's crust – these are very unreactive metals, such as gold and platinum. They are found as elements because they have never reacted with elements like oxygen.

Metals found in compounds

Very few metals occur as elements. The more reactive a metal is, the more easily it reacts with other substances to form compounds. Most metals are obtained from their compounds by chemical reactions. These compounds, which are mainly metal oxides, are found in rocks called **ores**. Ores are rocks that contain enough of a compound to extract a metal for profit. For example, the ore haematite contains enough iron oxide for it to be profitable to extract iron from it.

Bauxite (aluminium)   Cinnabar (mercury)   Galena (lead)   Haematite (iron)   Malachite (copper)

C *Metals are obtained from ores.*

Extracting metals from ores

The process of getting a metal from a compound in a rock by a chemical reaction is called **extraction**.

Some metals can be extracted by heating their compounds with carbon. For example, iron is extracted by heating iron oxide, from iron ores, with carbon. The word equation for this reaction is:

iron oxide + carbon → iron + carbon dioxide

Other metals, such as aluminium, are extracted from their compounds by electrolysis of a molten compound. Electrolysis decomposes some compounds into

1 a Name two metals found in the Earth's crust as elements.
b Why are these metals found as uncombined elements and not in compounds?

2 a What is an ore?
b What type of compounds are in most ores?

their elements. The ore bauxite contains aluminium oxide. The word equation for the electrolysis of aluminium oxide is:

aluminium oxide → aluminium + oxygen

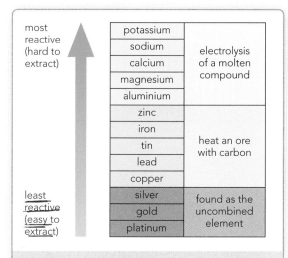

most reactive (hard to extract)	potassium	electrolysis of a molten compound
	sodium	
	calcium	
	magnesium	
	aluminium	
	zinc	heat an ore with carbon
	iron	
	tin	
	lead	
	copper	
least reactive (easy to extract)	silver	found as the uncombined element
	gold	
	platinum	

D *Reactivity series of metals*

The way in which a metal is extracted depends on its reactivity – the more reactive a metal is, the harder it is to extract. Figure D shows the **reactivity series**, which lists metals in order of reactivity. The most reactive metals are at the top.

The metals with very low reactivity can be found as elements. Metals up to and including zinc are extracted by heating with carbon. Metals above zinc in the reactivity series are extracted from their compounds by electrolysis of the molten compound.

ResultsPlus
Watch Out!

Most metals are not found as elements in ores. Instead, they have to be extracted by a chemical reaction from compounds found in those ores.

Skills spotlight

The higher the reactivity of a metal, the harder it is to extract from compounds found in ores. Use this idea to explain how the method of extraction for a metal can be predicted from its reactivity.

The harder a metal is to extract, the more it costs to extract. For example, electrolysis is more expensive than heating with carbon due to the cost of electricity. This means that the more reactive metal, the harder and more expensive it is to extract.

3 Give *two* common methods used to extract metals from ores.

4 Write a word equation for the extraction of tin by the reaction of tin oxide with carbon.

5 Write a word equation for the extraction of sodium by the electrolysis of molten sodium chloride.

6 Explain which method of extraction would be used for the following metals:
a calcium b lead c zinc.

H 7 a Write a balanced equation for the extraction of iron from iron oxide (Fe_2O_3) by heating with carbon. Include state symbols.
b Write a balanced equation for the extraction of aluminium from aluminium oxide (Al_2O_3) by electrolysis.

8 Explain why different metals are extracted in different ways. Give examples in your answer.

Learning Outcomes

4.1 Recall that:
a most metals are extracted from ores found in the Earth's crust
b unreactive metals are found in the Earth as the uncombined elements

4.2 Describe how most metals are extracted from their ores by: a heating with carbon, illustrated by iron b electrolysis, illustrated by aluminium (Knowledge of the blast furnace or the electrolytic cell for aluminium extraction are not required)

4.3 Explain why the method used to extract a metal is related to its position in the reactivity series and cost of the extraction process

HSW **3** Describe how phenomena are explained using scientific theories and ideas

C1.17 Metal extraction

Which metals can be extracted by heating with carbon?

The Stone Age came to an end with the discovery of how some metals could be extracted from their ores. The Bronze Age followed the Stone Age as people discovered how to extract copper and tin and combine them to make bronze. The Iron Age followed the Bronze Age once people had worked out how to extract the iron from its ore.

Metals and their ores are found in the Earth's crust. Metals which are lower in the reactivity series can be extracted from their ores by heating with carbon. Metals which are higher in the reactivity series cannot be extracted by heating with carbon and need to be extracted by electrolysis.

A The discovery of how to extract metals meant stronger and longer lasting tools could be made. This bronze arrowhead was found near Naples, Italy, and dates back to 1500 BC.

B This iron knife was developed once people learned to extract iron.

Your task

You are going to plan an investigation that will allow you to find out what happens when you heat some metal oxides with carbon. You are going to test the hypothesis that some metals can be extracted from their ores by heating with carbon. Your teacher will provide you with some materials to help you organise this task.

Learning Outcomes

4.4 Investigate methods for extracting a metal from its ore

Build Better Answers

When completing an investigation like this, one of the skills you will be assessed on is your ability to *plan a practical*. There are 4 marks available for this skill. Here are two student extracts that focus on this skill. Other skills that you need for the assessment are dealt with in other lessons.

Student extract 1 — A basic response for this skill

In order to carry out this investigation I am going to follow these steps: Mix a measured amount of a metal oxide in a crucible with a measured amount of powdered charcoal and put a thin layer of charcoal over the top. Heat the crucible with a Bunsen burner for five minutes. Let the crucible cool and have a look for any signs of a metal. Pour the contents of the crucible into a beaker of water and look for any metal which might sink to the bottom. I am going to test three different metal oxides in this way.

> A plan should be clear like this with logical steps so that someone else could easily follow your instructions.

> It is important to say what is going to be tested or measured.

Student extract 2 — A good response for this skill

> The information from the first student extract would be included here as well.

My hypothesis is that some metals can be extracted from their ores by heating with carbon. If I carry out the investigation in this way I will be able to test several different metal ores and see if a metal is formed when they are reacted with carbon.

I am going to test magnesium oxide, iron oxide and copper oxide. I am going to use these oxides because they will give me a range of different metals to test the hypothesis with. Some of these metals are high up in the reactivity series and some are lower down.

> The investigation needs to explain how it will test the hypothesis.

> Give a range of tests which will give a good set of data and explain why you have chosen this range.

ResultsPlus

To access 2 marks

- Provide a logically ordered method
- Choose a range of data or observations that will test the hypothesis

To access 4 marks

You also need to:

- Explain how your method will test the hypothesis
- Explain why you have chosen your range of data or observations

What are oxidation and reduction?

Part of the Swedish town of Kiruna is being gradually moved four kilometres further north. Many homes and buildings will be demolished, but others will be loaded onto trailers and moved. The move is necessary to reduce the risk of buildings collapsing into the iron ore mine that has been spreading beneath the town.

A *The town of Kiruna is being moved.*

> **?**
> 1 a Define oxidation.
> b Define reduction.
>
> 2 Decide whether each of the following is oxidation or reduction:
> a the addition of oxygen
> b the loss of oxygen
> c the extraction of nickel from nickel oxide
> d the corrosion of nickel to form nickel oxide
> e the iron in a nail being converted to rust.

Metal extraction is reduction

Most of the compounds found in ores from which we extract metals are metal oxides. In order to obtain metals from their oxides, the oxygen must be removed. When oxygen is removed from a compound it is said to be reduced – so, the extraction of metals from their oxides is **reduction**.

Iron is obtained by the removal of oxygen from iron oxide by heating with carbon. The iron oxide is reduced to iron:

iron oxide + carbon → iron + carbon dioxide

B *Molten aluminium oxide is reduced to aluminium by electrolysis.*

Aluminium is obtained by the removal of oxygen from aluminium oxide by electrolysis. The aluminium oxide is reduced to aluminium:

aluminium oxide → aluminium + oxygen

Corrosion of metals is oxidation

Most metals corrode. The **corrosion** of iron is called **rusting**. Corrosion happens when the surface of a metal changes by reaction with oxygen – sometimes with water as well. **Oxidation** is the addition of oxygen to a substance. When metals corrode the metal is oxidised.

The more reactive the metal, the more rapidly it corrodes. Metals with low reactivity are more resistant to oxidation and corrosion. For example, gold does not corrode at all.

Reactive aluminium does not corrode as much as expected because its surface oxidises quickly to form a protective layer of aluminium oxide (Al_2O_3) – this stops any further corrosion.

C *The Angel of the North is made of iron, so it is corroded by oxygen in the air. The corrosion of iron is oxidation.*

3 Which type of metals does not corrode?

4 Three metals were left outside for one month. Metal A had not corroded at all; metal B had corroded a lot; metal C had corroded a little. Put the metals in order of reactivity and explain your reasoning.

5 Aluminium is a reactive metal. However it is resistant to corrosion.
a Explain why the resistance of aluminium to corrosion is somewhat surprising.
b Suggest why aluminium is resistant to corrosion.

Ⓗ 6 a Write a balanced equation for the reduction of zinc oxide (ZnO) to zinc by carbon.
b Write a balanced equation for the oxidation of zinc to zinc oxide (ZnO).

7 Describe what oxidation and reduction are. Use your answer to explain how oxidation and reduction are involved in the extraction and corrosion of metals. Include examples of word equations.

Learning Outcomes

4.5 Describe oxidation as the gain of oxygen and reduction as the loss of oxygen

4.6 Recall that the extraction of metals involves reduction of ores

4.7 Recall that the oxidation of metals results in corrosion

4.8 Demonstrate an understanding that a metal's resistance to oxidation is related to its position in the reactivity series

HSW 3 Describe how phenomena are explained using scientific theories and ideas

Can we make gold from rubbish?

Why is recycling useful?

A tonne of waste electronic and electrical equipment (WEEE) contains more gold than 4 tonnes of gold ore. However in the UK we throw away about 1 million tonnes of this waste every year, much of it ending up in landfill sites. So now there are plans to start mining gold and other metals from buried waste in old landfill sites.

A The WEEE man is constructed from 3.3 tonnes of electrical and electronic equipment that the average person in the UK throws away in a lifetime.

Recycling metals

Many metals can be **recycled**. This is when used metal is melted down and made into something new. These are some of the main advantages of recycling.

B Huge brick pipes (flues) used to carry poisonous sulfur dioxide away from lead mines in Yorkshire

- Natural reserves of metal ores will last longer if more metals are recycled.
- For most metals, less energy is needed to recycle them than is needed to extract them from their ores. For example, recycling aluminium can use just 5% of the energy needed to extract aluminium from its ore. This energy saving also makes recycled metals cheaper.
- Recycling reduces the need to mine ores. Mining can damage the landscape and create noise and dust pollution in the same way as limestone quarrying (see C1.6).
- Recycling produces less pollution. For example, sulfur dioxide is formed when lead is extracted from the ore galena, which contains lead sulfide, and carbon dioxide, a greenhouse gas, is emitted when fossil fuels are used to generate the energy for electrolysis or reduction of ores.
- More recycling means that less waste metals are disposed of in landfill sites.

However, there are also some disadvantages of recycling – including the costs and the energy used in collecting, sorting and transporting metals to be recycled. At times, for some metals, it can be more expensive to recycle.

1 What do we mean by *recycling* metals?

2 Make a table with two columns – one showing advantages and one showing disadvantages of recycling metals compared to extracting more of the metals from ores.

How metals are recycled

The first stage of recycling metals involves collecting the metals, for example taking used batteries to collection points in shops. However, for this to work people have to be prepared to separate their rubbish.

Iron and steel are easily separated from other metals because they are magnetic. Other metals are usually separated by hand. The metals obtained from waste are often melted down to form blocks and sold on to manufacturers.

C *Aluminium cans and steel cans are easily separated.*

D *These ingots are made from recycled aluminium drink cans.*

Figure E shows what percentage of metals used today are recycled.

Metal	% of metal used that is recycled
aluminium	40
copper	30
lead	75
iron (and steel)	40

E *Use of recycled metals in the UK*

Skills spotlight

Explain why the government sets targets for local councils to increase metal recycling over the next few years. Hint: think about economic and environmental factors.

3 Draw a bar chart using the data in Figure E to show what percentage of different metals used is recycled.

4 If you were on your school council, what could you do to increase the amount of metals recycled at your school?

5 How are iron and steel separated from other metals for recycling?

6 How is aluminium recycled?

7 If something is done in a sustainable way, it allows things to be available for future generations. Explain why recycling metals is more sustainable than extracting metals from ores.

Results Plus
Watch Out!

When discussing recycling, make sure that you take into account both economic and environmental effects.

Learning Outcomes

4.9 Discuss the advantages of recycling metals, including economic implications and how recycling preserves both the environment and the supply of valuable raw materials

HSW 13 Describe the social, economic and environmental effects of decisions about the uses of science and technology

C1.20 Properties of metals

What are metals used for?

In 1779, the world's first cast iron bridge was built across the River Severn at Coalbrookdale, allowing easier transport across the river. The use of iron as a structural material revolutionised bridge building allowing better bridges to be built than previously. Coalbrookdale also gave the world its first iron rails and iron wheels.

A The world's first iron bridge

> **1** Which of the following are properties of metals: electrical conductor; low melting point and low boiling point; malleable; strong; thermal insulator; shiny?

Properties of metals

Metals have many useful properties. For example, they are shiny when polished, they **conduct** heat and electricity, they are **malleable** (can be hammered into shape) and they are **ductile** (can be stretched into wires). Different metals have slightly different properties, making them useful for different things.

ResultsPlus
Watch Out!

When asked to explain the uses of a specific metal, link its properties to its uses. For example, copper is a good conductor of electricity so it is used for wires.

B Aluminium is used in the manufacture of passenger aircraft because of its low density.

Aluminium

Aluminium is a metal with a low **density**. It does not corrode because it has a layer of aluminium oxide that forms quickly on its surface. Aluminium is used, with other metals, to make aeroplanes. The lighter an aeroplane is, the less fuel it will need as it flies. Many cars also have bodies that contain a lot of aluminium to reduce their weight.

> **2** Explain why aluminium is used for making aircraft.
>
> **3** Explain why copper is used for making water pipes.
>
> **4** Explain why copper is used for making electrical cables.

Copper

Copper is an extremely good electrical conductor and so electrical cables are made using copper. Copper has low reactivity and does not react with water – this makes it ideal for use in water pipes. Lead used to be used in water pipes, but it slowly reacted with water and could cause lead poisoning.

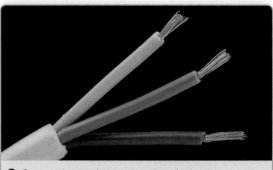

C Copper is used to make electrical cables.

Gold

Gold is a very unreactive metal and does not corrode. It is also very attractive. Its main use is for jewellery because it can be worked easily into beautiful shapes and the metal remains shiny because it does not tarnish from corrosion. Gold is also one of the best electrical conductors of all the metals and is used inside most electronic devices, including mobile phones and computers. It is used in printed circuit boards, connection strips and contacts in switches. Only a tiny amount is needed for these applications.

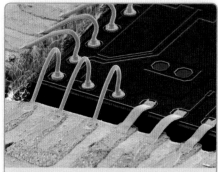

D There is a thin layer of gold on top of other conductors on this circuit board.

Iron and steel

Iron is used more than any other metal. This is because it is fairly cheap to extract from iron ore by heating with carbon. Pure iron is too soft to be very useful, so it is made into steel, which is stronger and harder. Steels are mixtures of iron with carbon and often other metals. Steel is used to make a huge range of things, including bridges, cars, electrical goods, machinery and the frames for buildings. It is also magnetic, which can be useful. However, both iron and steel rust because the iron reacts with air and water.

E Starting in the 1880s, the development of steel-framed buildings led to the construction of skyscrapers. This photo was taken in New York in 1932.

Skills spotlight

Think of a hypothesis that compares the properties of two of the metals mentioned in this lesson. How would you test your hypothesis?

5 Explain why gold is used for making jewellery.

6 Explain why iron, in the form of steel, is used for making the frames for buildings.

7 Explain why many modern cars have bodies that contain a lot of aluminium.

8 Ships are made out of steel. Give some advantages and disadvantages of making them out of aluminium instead.

9 Describe some of the benefits that metals have brought to modern life.

Learning Outcomes

4.10 Describe the uses of metals in relation to their properties, including: a aluminium b copper c gold d steel

HSW 5 Plan to test a scientific idea, answer a scientific question, or solve a scientific problem by selecting appropriate data to test a hypothesis

Can we alter the properties of metals?

How can we improve the properties of metals?

A big problem with iron is that it rusts in moist air. However, iron can be made into stainless steel which does not rust. This huge arch in St. Louis, USA, is made of stainless steel and will not rust.

A *Stainless steel Gateway Arch St. Louis, Missouri, USA*

Many metals are mixed with small amounts of other metals to improve their properties for a specific application. A mixture like this is called an **alloy**.

In a pure metal structure, all the atoms are the same size and they are packed closely together. This means that layers of atoms can slide over each other, which makes the metal soft. In an alloy, there are some atoms of other elements that are different sizes. This prevents atoms from sliding past each other so easily and makes alloys harder and stronger.

> **ResultsPlus**
> **Watch Out!**
>
> An alloy is a mixture of metals, and the alloy is often harder and stronger than original metals.

?

1 What is an alloy?

2 Why are pure metals often not very useful? Explain your answer using the idea of particles.

3 a What is an alloy steel?
b Why is an alloy steel stronger than iron? Explain your answer using the idea of particles.

4 a Suggest a use for a steel alloy containing chromium and nickel.
b Why is this alloy well suited to this use?

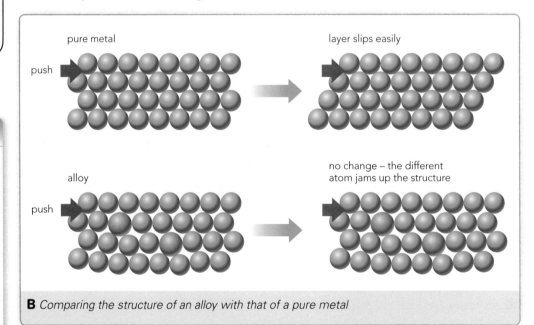

B *Comparing the structure of an alloy with that of a pure metal*

Examples of alloys

There are many different **alloy steels** in which small amounts of different metals are mixed with iron. Alloy steels are stronger than iron (they need a bigger force to break them) and some resist corrosion. Stainless steel, an alloy of iron with small amounts of chromium and nickel, does not corrode at all.

H

Pure gold is too soft to be used successfully in jewellery. Other metals, such as copper and silver, are added to it to make a harder and stronger alloy. The purity of gold is measured in **carats** or as a **fineness**. Pure gold is 24-carat and has a fineness of 1000 parts per thousand. Figure C shows how the carat value and fineness change as more of other metals are added.

D *Gold alloys vary in the proportion of gold they contain.*

Percentage gold	Percentage other metals	Carats	Fineness (parts per thousand)
100%	0%	24	1000
91.7%	8.3%	22	917
75%	25%	18	750
50%	50%	12	500
37.5%	62.5%	9	375

C *Purity of gold*

Shape memory alloys

Nitinol is an alloy of nickel and titanium. It is a **smart material** – these materials have a property that changes with a change in conditions, usually temperature. Nitinol is a **shape memory alloy**. If the shape of something made of nitinol is changed, it returns to its original shape when heated. One example of the use of nitinol is the repair of a collapsed artery. Doctors slide a cooled and squashed nitinol tube into the damaged artery, and as it warms up in the body it returns to its original size, this holds the artery open. Other uses of shape memory alloys include flexible spectacle frames.

Skills spotlight

H Scientists often need to present information in mathematical ways that are different from everyday use. A jeweller is selling a ring that has a fineness of 800. How could he explain what this means to a customer?

H 5 a Why is jewellery not made out of pure gold? **b** Which contains more gold – 20-carat gold or 9-carat gold? **c** Which contains more gold – 900 or 850 fineness? **d** Which contains more gold – gold with a fineness of 600 or 14-carat gold?

H 6 a Derive an equation that converts fineness of gold into carats. **b** Plot a graph to show the relationship between carats and fineness.

H 7 a What is a smart material? **b** Name a smart material that is an alloy.

8 Explain what an alloy is and why they are often stronger than pure metals. Include a diagram in your answer.

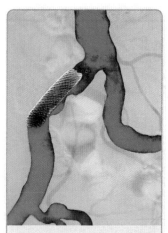

E *Using a shape memory alloy tube to open up an artery*

Learning Outcomes

4.11 Use models to explain why converting pure metals into alloys often increases the strength of the product

4.12 Demonstrate an understanding that iron is alloyed with other metals to produce alloy steels with a higher strength and a greater resistance to corrosion

H 4.13 Describe how alloying changes the properties of metals, including: **a** smart or shape memory alloys, including nitinol, an alloy of nickel and titanium **b** gold alloys with higher strength, including fineness (parts per thousand) and carats to indicate the proportion of pure gold

H 4.14 Demonstrate an understanding that new materials are developed by chemists to fit new applications, such as the creation of new shape memory alloys for use, for example, in spectacle frames and as stents in damaged blood vessels

HSW 11 Present information, develop an argument and draw a conclusion, using scientific, technical and mathematical language, and ICT tools

What is crude oil used for?

The sand and clay deposits near the Athabasca River in Canada contain an extremely sticky form of crude oil. 'Oil sands' such as this may contain over half of the world's remaining reserves of oil. But is it economic to mine these sands – and is it worth the damage to habitats?

A *Bitumen-soaked sand can be dug out of the ground at oil sands in Alberta, Canada.*

Petrol and other useful products are obtained from **crude oil**. Crude oil is a thick, black liquid found in some sedimentary rocks. Crude oil and natural gas are normally found deep underground, trapped by layers of rock that prevent the liquid and gas from rising.

Crude oil and natural gas are **fossil fuels**. They formed over millions of years from the remains of microscopic plants and animals that once lived in the sea. When these died, they fell to the sea bed and were then buried by sediments. The sediments kept oxygen away and stopped the remains decaying. As more sediments built up on top of the remains, the heat and pressure increased and they gradually turned into oil or gas.

B *Crude oil is a liquid.*

1 How is crude oil formed?

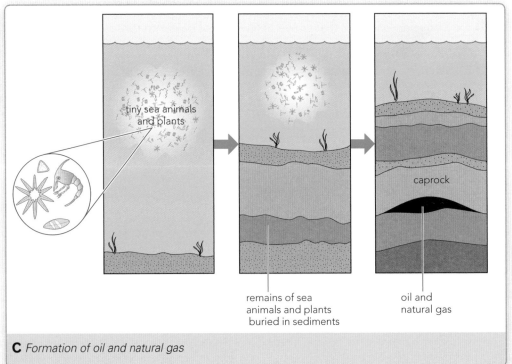

tiny sea animals and plants

caprock

remains of sea animals and plants buried in sediments

oil and natural gas

C *Formation of oil and natural gas*

Crude oil is a **mixture** of different **hydrocarbon molecules**. A hydrocarbon is a compound that contains only carbon and hydrogen atoms. The different hydrocarbons in crude oil have different numbers of carbon atoms in their molecules, ranging from one carbon to many hundreds. Crude oil can also contain some impurities like sulfur.

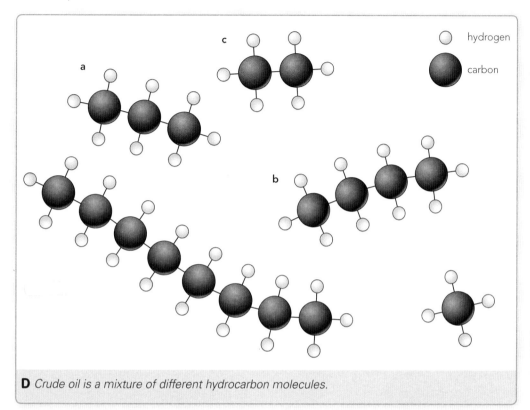

D *Crude oil is a mixture of different hydrocarbon molecules.*

hydrogen

carbon

Crude oil is a **non-renewable resource**. It is probably being formed somewhere under the ground right now, but we are pumping oil out of the ground much faster than it is being made. It is very difficult to estimate the amount of oil remaining in the ground (and how much of it can be extracted economically) but at the present rate of consumption we probably have about 40–50 years' supply left. At present, most oil is obtained by drilling wells. However, as the oil from wells runs out and as the demand for oil increases, surface mining of oil sands (Figure A) will increase.

4 Why is crude oil described as a 'non-renewable' resource?

5 Look at molecules a, b and c in Figure D. How are these molecules different?

6 Explain how the energy stored in crude oil originally came from the Sun.

2 a What is a hydrocarbon?
b What can crude oil contain in addition to hydrocarbons?

3 Why is crude oil described as a mixture?

Results Plus
Watch Out!

Hydrocarbons are *compounds* of hydrogen and carbon. Crude oil is a *mixture* of different hydrocarbon compounds.

Skills spotlight

People who make decisions about scientific issues often need to consider many factors. Should we continue to extract oil, even if it causes environmental damage? What would you need to know to make an informed decision?

Learning Outcomes

5.1 Describe hydrocarbons as compounds that contain carbon and hydrogen only

5.2 Describe crude oil as a complex mixture of hydrocarbons

HSW **13** Describe the social, economic and environmental effects of decisions about the uses of science and technology

C1.23 Crude oil fractions

How are the different hydrocarbons in crude oil separated?

The black liquid in Figure A is bitumen. It is being used to stick the roofing felt down to make it waterproof. Both bitumen and petrol are obtained from crude oil, so how are the liquids separated?

A *Bitumen being used as part of a waterproof roof*

Skills spotlight

Think up one question about crude oil that cannot be answered by science and one question that can.

The hydrocarbons in crude oil are not easy to use when they are all mixed together. Crude oil is sent to an oil refinery for the mixture to be separated into simpler mixtures. This is done by **fractional distillation**. Distillation is a process used to separate mixtures of different liquids. The mixture of liquids is boiled and the vapour from it is condensed. The simpler mixtures produced by the fractional distillation of crude oil are called **fractions**.

1 a What is a 'fraction' of crude oil?
b What is the name of the process used to separate the fractions?

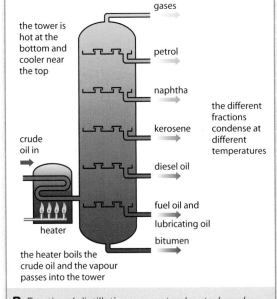

B *Fractional distillation separates heated crude oil into fractions.*

- gases
- the tower is hot at the bottom and cooler near the top
- petrol
- naphtha
- kerosene
- crude oil in
- the different fractions condense at different temperatures
- diesel oil
- fuel oil and lubricating oil
- heater
- bitumen
- the heater boils the crude oil and the vapour passes into the tower

C *Fractional distillation happens in tall columns in an oil refinery.*

ResultsPlus
Watch Out!

All fractions are mixtures of compounds. Fractions have different properties – see Figure E.

Each different fraction of the crude oil is still a mixture of different hydrocarbon molecules, but each fraction contains a group of hydrocarbons with a more limited range of numbers of carbon atoms in a molecule. For example, all the hydrocarbons in the petrol fraction have 5, 6 or 7 carbon atoms per molecule.

Crude oils obtained from different parts of the world contain different percentages of the different fractions. This makes some crude oils more valuable than others.

The different fractions have different uses because they have different properties.

Fraction	Uses
gases	• fuel for vehicles (liquid petroleum gas, or LPG) • bottled gas for camping stoves • heating and cooking in homes
petrol	fuel for cars
kerosene	fuel for aircraft engines
diesel oil	fuel for diesel engines (some cars, lorries, trains)
fuel oil	• fuel for large ships and some power stations • fuel for heating • lubricating oil
bitumen	making roads, waterproofing flat roofs

D *Uses of the different fractions in crude oil*

2 a Write down the names of two fractions in Figure D that are not only used as fuels.
b Give another use of each of these fractions.

3 How many carbon atoms might there be in the carbon chain of a molecule of:
a a hydrocarbon in the gases fraction
b petrol
c bitumen?

Fraction	Length of molecule	Ease of ignition	Boiling point	Viscosity
gases	short carbon chains (only a few carbon atoms) ↓ long carbon chains (up to 40 Carbon atoms)	easy ↓ difficult	low (< 0°C) ↓ high (> 350°C)	runny ↓ thick and sticky
petrol				
kerosene				
diesel oil				
fuel oil				
bitumen				

E *Properties of the molecules in the different fractions*

The fractions coming from the top of the column contain short chain molecules, so they have low **boiling points** and **ignite** (set alight) easily. They have low **viscosity** (they are runny) when they are liquids. The fractions from the bottom of the column have much longer chain molecules and have higher boiling points and are harder to ignite. They are viscous (thick and sticky) liquids.

4 What is the link between the number of atoms in the carbon chain of a hydrocarbon molecule and: a how easy it is to ignite b its boiling point c its viscosity?

5 What are the differences between the fuel oil used for ships and the kerosene used in jet engines?

6 Explain why crude oil is separated into different fractions.

Learning Outcomes

5.3 Describe the separation of crude oil into simpler, more useful mixtures by the process of fractional distillation (details of fractional distillation are not required)

5.4 Recall the name and uses of the following fractions: a gases, used in domestic heating and cooking b petrol, used as a fuel for cars c kerosene, used as a fuel for aircraft d diesel oil, used as fuel for some cars and trains e fuel oil, used as fuel for large ships and in some power stations f bitumen, used to surface roads and roofs

5.5 Describe that hydrocarbons in different fractions differ from each other in:
a the number of carbon and hydrogen atoms their molecules contain b boiling points c ease of ignition d viscosity

HSW 4 Identify questions that science cannot currently answer, and explain why these questions cannot be answered

How does burning help jet fighters?

C1.24 Combustion

What is produced when hydrocarbons burn?

This jet fighter uses kerosene as fuel in its engine. When it needs a lot of power, extra kerosene is injected into the engine's exhaust gases (in the after burner) where it burns. This can nearly double the force obtained from the engine.

A *Eurofighter with its after burners on*

1 Give an example of a hydrocarbon fuel.

2 Why is combustion an example of an oxidation reaction?

3 Look at Figure B. What is the source of oxygen that allows the candle to burn?

4 In Figure B, which part of the apparatus shows that:
a carbon dioxide is produced
b oxygen is produced?

When hydrocarbon fuels burn they react with oxygen and release heat and light energy. This is an oxidation reaction called **combustion**.

The apparatus in Figure B can be used to detect the products of a combustion reaction. The U-tube in the middle contains anhydrous copper sulfate. This white substance turns blue when water is added. **Limewater** is used to test for carbon dioxide gas – it turns milky if carbon dioxide is bubbled through it.

When this apparatus is used to test burning hydrocarbons, the anhydrous copper sulfate turns blue and the limewater turns milky. This shows that water and carbon dioxide are present in the waste gases.

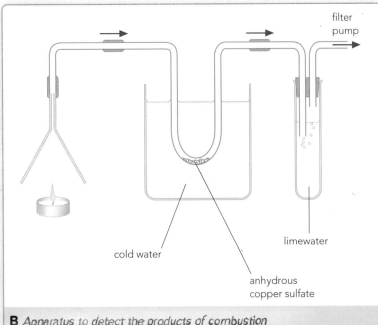

filter pump

cold water

limewater

anhydrous copper sulfate

B *Apparatus to detect the products of combustion*

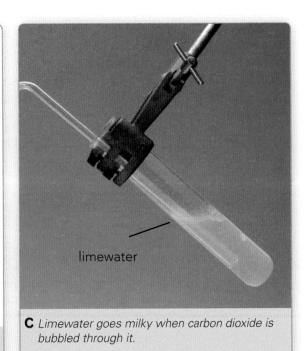

limewater

C *Limewater goes milky when carbon dioxide is bubbled through it.*

All hydrocarbon fuels produce carbon dioxide and water when they burn, as long as enough oxygen is present. Figure D shows what happens when methane burns. Methane is the main gas in natural gas.

The word equation and the symbol equation for the reaction when methane burns are:

methane   +   oxygen   →   carbon dioxide   +   water

H CH_4 + $2O_2$ → CO_2 + $2H_2O$

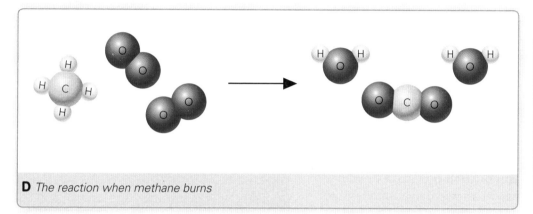

D *The reaction when methane burns*

In the combustion of a hydrocarbon, the reaction is described as **complete combustion** if all the hydrocarbon is used up and the only products are carbon dioxide and water.

5 Butane is a hydrocarbon compound found in crude oil. Write a word equation for the combustion of butane.

6 Limewater is calcium hydroxide solution, $Ca(OH)_2$. It forms calcium carbonate, $CaCO_3$, and water when it reacts with carbon dioxide.
a Write a word equation for this reaction.
b Why does limewater go milky when carbon dioxide is bubbled through it?
H c Write a balanced equation for the reaction.

7 Jose says, 'Burning petrol produces nitrogen oxides and water'. Explain how you can tell he is wrong just by using the fact that petrol is a mixture of hydrocarbon molecules.

Results Plus
Watch Out!

The products of complete combustion of a hydrocarbon are always carbon dioxide and water – nothing else.

Skills spotlight

Scientists use various tests to identify the products of reactions. If you have a sample of gas that you think may be carbon dioxide, how could you find out if you are right?

Learning Outcomes

5.6 Describe how the complete combustion of hydrocarbons:
a involves oxidation of the hydrocarbons
b produces carbon dioxide and water
c gives out energy

5.7 Describe the chemical test for carbon dioxide (using limewater)

HSW 1 Explain how scientific data is collected and analysed

What happens if burning fuel does not get enough oxygen?

Figure A shows an oil-storage depot which caught fire. The fire burned for several days, and smoke covered a large part of south-east England. The black smoke shows that the fuel was not burning completely.

A *The fire at Buncefield, in the south of England*

Watch Out!

Carbon monoxide is a product of incomplete combustion, but carbon dioxide and water are also produced.

When a hydrocarbon fuel burns with as much oxygen as it needs, it produces only carbon dioxide and water. However, sometimes a burning fuel may not have enough oxygen – in this case **incomplete combustion** occurs. The hydrogen atoms form water, but there is not enough oxygen for all the carbon atoms to form carbon dioxide. Some of the carbon may form **carbon monoxide**, and some may just form solid particles of carbon (**soot**).

The word equations show what can happen if methane burns without enough oxygen.

methane + oxygen → carbon + water
methane + oxygen → carbon monoxide + water
methane + oxygen → carbon dioxide + carbon monoxide + carbon (soot) + water

1 Write a word equation for:
a the complete combustion of methane
b the incomplete combustion of methane forming soot.

Different percentages of carbon, carbon monoxide and carbon dioxide are produced, depending on the amount of oxygen available.

The carbon appears as tiny particles that you can see as black smoke in a bonfire. The yellow colour of a flame in Figure B is caused by the hot carbon particles (soot) glowing. The soot produced by faulty gas fires or boilers makes marks on the walls of a room.

B *A Bunsen burner has a blue flame when complete combustion is occurring. When the air hole is closed, incomplete combustion occurs and the flame is yellow.*

Carbon monoxide problems

Carbon monoxide is an odourless, colourless toxic gas. It reduces the amount of oxygen that can be carried around the body in the blood. Around 40 people die each year in the UK from carbon monoxide poisoning due to faulty gas boilers and fires. Hundreds more have to be taken to hospital. Figure C shows just one way in which this can happen. Carbon monoxide is also produced by car engines.

Harm from carbon monoxide in our homes can be reduced by:

- making sure that all fuel-burning appliances are serviced regularly
- fitting homes with carbon monoxide detectors, to warn people if an appliance has stopped working properly.

Soot problems

Soot produced in appliances such as boilers can clog up the pipes carrying the waste gases away. It can even cause fires in chimneys if enough builds up.

Soot is also produced by vehicles, especially vehicles that use diesel fuel. Small particles of soot can collect in the lungs when people breathe in sooty air, and this can cause lung diseases. Soot also makes buildings dirty.

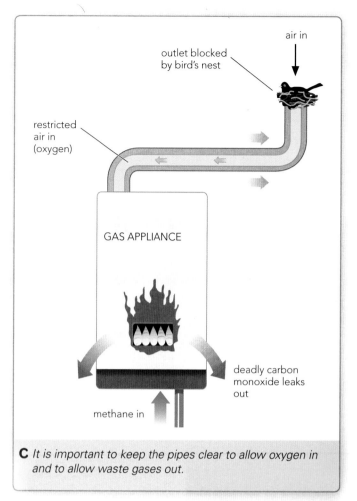

C It is important to keep the pipes clear to allow oxygen in and to allow waste gases out.

2 What toxic gas is produced during the incomplete combustion of a hydrocarbon?

3 In a Bunsen burner:
a How do you get a blue flame?
b Why is the blue flame sometimes called a 'clean' flame?
c What causes the yellow colour in the flame when the air hole is closed?
d Why does this happen?

4 Why is it important to have gas boilers serviced regularly?

5 Describe two problems that soot can cause.

6 You notice black marks on the wall above the boiler in your home. Why should you stop using the boiler immediately? Explain in as much detail as you can.

Skills spotlight

Scientists need to be able to communicate ideas clearly to reach everybody. Design a poster to warn people about the dangers of carbon monoxide poisoning and how to prevent it.

Learning Outcomes

5.8 Explain why the incomplete combustion of hydrocarbons can produce carbon and carbon monoxide

5.9 Describe how carbon monoxide behaves as a toxic gas

5.10 Demonstrate an understanding of the problems caused by incomplete combustion producing carbon monoxide and soot in appliances that use carbon compounds as fuels

HSW 11 Present information, develop an argument and draw a conclusion, using scientific, technical and mathematical language, and ICT tools

How can products of burning fuels kill fish?

What is acid rain and what harm does it do?

This helicopter is adding lime to a lake in Norway. This is to neutralise the waters in the lake, which have become too acidic for fish to live in.

A Neutralising a lake

Discovering the problem

In the 1970s, the numbers of fish caught in rivers and lakes in southern Norway started to decrease. Norwegian scientists noticed that these lakes and rivers were much more acidic than those in other parts of the country that still had healthy fish. Trees in the area were also being damaged.

The scientists looked at weather patterns and noticed that the winds blowing over these areas had passed over western and central Europe. They said that pollution from factories and power stations in these parts of Europe was being carried in the atmosphere and was making the rainfall acidic – this acidic rain was killing the fish. Similar problems were being found in lakes in Canada, caused by pollution from factories in the USA.

Scientists and politicians in other countries suggested other possible causes for the acidified lakes and rivers. In the 1980s, a research programme was carried out into the problem. This confirmed that pollution was responsible for the acidic lakes and the deaths of the organisms living in those lakes.

B These trees have been damaged by acid rain.

> 1 Why is rain normally slightly acidic?
>
> 2 a What is 'acid rain'?
> b What causes acid rain?

The causes of acid rain

Rainwater is naturally slightly acidic because carbon dioxide and other acidic gases in the air dissolve in it. **Acid rain** is rain that is more acidic than normal and has a pH lower than 5.2.

Hydrocarbon fuels (such as petrol and coal) do not just contain hydrocarbons but also impurities like sulfur. When the fuel is burnt, the sulfur reacts with oxygen from the air to form sulfur dioxide gas. This dissolves in rainwater, making it more acidic (lower pH).

Effects and solutions

The effects of acid rain include:
- making rivers, lakes and soils acidic – this harms organisms living there
- damaging trees
- speeding up the weathering of buildings made of limestone or marble and the corrosion of metal.

The problem of acid rain in Europe and North America is being solved by reducing the amount of sulfur in petrol, diesel and fuel oil and removing acidic gases from power station emissions (see C1.9). However, it is still a problem in other parts of the world where industries are still developing.

C *This statue of the Buddha in China is being damaged by acid rain caused by nearby coal-fired power stations.*

Skills spotlight

Decisions about science and technology need to be made by scientists and governments, often working together. Suggest why the UK government decided to support the introduction of low-sulfur fuels and the installation of equipment to remove sulfur dioxide from power station chimneys.

Results Plus
Watch Out!

All rainwater is slightly acidic naturally. Acid rain is more acidic than normal due to sulfur dioxide dissolving in it.

3 What effect does acid rain have on:
a the water in rivers and lakes
b organisms that live in rivers and lakes
c trees
d stone buildings and statues?

4 Limestone (calcium carbonate) can be used to neutralise the sulfuric acid that makes lakes acidic. Write a word equation for this neutralisation reaction.

5 Describe how acid rain forms and describe two ways in which acid rain in Europe has been reduced.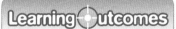

Learning Outcomes

5.11 Explain why impurities in some hydrocarbon fuels result in the production of sulfur dioxide

5.12 Demonstrate an understanding of some problems associated with acid rain caused when sulfur dioxide dissolves in rain water

HSW 13 Explain how and why decisions that raise ethical issues about uses of science and technology are made

>>>>>>>>>>>>>>>>>>>>>>>>> Is having carbon dioxide in the air a good thing?

How can human activities affect the Earth's temperature?

Many people think that adding carbon dioxide to the atmosphere by burning fossil fuels is helping to cause global warming, which is causing rising sea levels. The Maldives are islands in the Indian Ocean. Their mean height above sea level is only about 2 metres. In 2009, the Maldives government held an underwater meeting, to draw attention to the danger to their country!

A *Underwater government meeting in the Maldives*

ResultsPlus
Watch Out!

Carbon dioxide forms a very small proportion of the atmosphere. Even after the increase in burning fossil fuels during the last 200–300 years, the amount of carbon dioxide in the atmosphere is less than 0.4%.

The surface temperature of the Earth varies between the Equator and the poles and also because of changing seasons and weather patterns. The mean temperature can also change over long periods of time i.e. hundreds of years (as shown in Figure B).

Keeping the Earth warm

The mean surface temperature of the Earth now is about 14 °C. Some gases in the atmosphere, such as carbon dioxide, methane and water vapour, trap heat energy and help to keep the Earth warm. This is known as the greenhouse effect and these gases are often called greenhouse gases. Without these gases, the mean surface temperature of the Earth would be about −18 °C.

1 Look at Figure B. How has the mean world temperature changed since 1860?

The concentration of carbon dioxide in the atmosphere can change due to natural processes. However, the large increase in concentration since about 1800 is believed to be due to human activities, such as burning fossil fuels. The levels of other greenhouse gases are also thought to be increasing because of human activities such as farming.

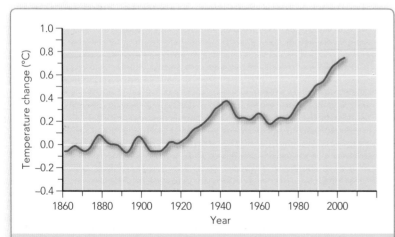

B *Change in mean world temperature from 1860. The temperatures are compared to the mean temperature in 1900.*

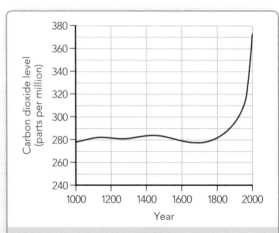

C *Change in concentration of carbon dioxide in the atmosphere (information obtained from air bubbles trapped in ice)*

The Intergovernmental Panel on Climate Change (IPCC) is a large group of scientists from all over the world. They report on research about changes to the world's climate and the possible causes and impacts. They have concluded that greenhouse gases emitted by human activities are causing an increased warming effect. An increase in the Earth's mean temperature of 4 to 7 °C over the next century is likely to change global weather patterns, causing climate change. Their conclusions are accepted by most scientists and governments around the world. However, not all scientists are convinced that human actions are causing climate change.

Reducing the amount of carbon dioxide

We can reduce the amount of carbon dioxide that human activities are adding to the atmosphere by reducing our use of fossil fuels. Chemists are investigating ways of controlling the amount of carbon dioxide in the atmosphere.

Two of the methods currently being investigated are:

- adding iron compounds to the oceans. This is known as iron seeding. Iron is an essential nutrient for plant growth and is often in short supply. Adding iron compounds encourages microscopic plants to grow. These use carbon dioxide as they photosynthesise and are then eaten by sea creatures. The idea is that when these die and sink to the ocean floor, the carbonate in their shells is buried. This removes the carbon from the atmosphere for a long time.

- capturing carbon dioxide from fossil-fuelled power stations and reacting it to make hydrocarbon compounds such as propane and butane. These hydrocarbons can be used as fuels.

> 2 Look at Figures B and C.
> a How has the amount of carbon dioxide in the atmosphere changed since 1800?
> b How does this compare to the change in mean world temperature?
>
> 3 Explain why your answers to question 2 do not show that changes in carbon dioxide levels cause the change in temperature.

> 4 Name two sources of:
> a carbon dioxide b methane.
>
> 5 Draw a flow chart to explain how iron seeding can reduce the amount of carbon dioxide in the atmosphere.
>
> 6 Describe one other way of reducing the amount of carbon dioxide in the atmosphere.
>
> 7 Hydrocarbons produced from carbon dioxide could be used as fuels. Explain whether or not you think this will help to reduce the amount of carbon dioxide in the atmosphere.

Skills spotlight

Scientists validate the work of other scientists by the process of 'peer review'. The IPCC has been criticised for not always using peer-reviewed evidence in its studies. Explain what peer review means and why peer-reviewed evidence is the best to use.

Learning Outcomes

5.13 Describe how various gases in the atmosphere, including carbon dioxide, methane and water vapour, trap heat from the Sun and that this keeps the Earth warm

5.14 Demonstrate an understanding that the Earth's temperature varies and that human activity may influence this

5.15 Demonstrate an understanding that the proportion of carbon dioxide in the atmosphere varies, due to human activity, and that chemists are investigating methods to control the amount of the gas in the atmosphere by: a iron seeding of oceans b converting carbon dioxide into hydrocarbons

5.16 Evaluate how far the correlation between global temperature and the proportion of carbon dioxide in the atmosphere provides evidence for climate change

HSW 14 Describe how scientists share data and discuss new ideas, and how over time this process helps to reduce uncertainties and revise scientific theories

C1.28 Biofuels

What are the advantages and disadvantages of biofuels?

The petrol used in cars in the UK mainly contains hydrocarbons. However, it also contains up to 5% ethanol – the same substance as in alcoholic drinks.

A Using plants to make biofuel means there are fewer plants used for growing food.

Biofuels are obtained from living organisms or organisms that have recently died. Biofuels include wood and dried animal droppings, which have been used as fuels for hundreds of thousands of years and are still important as fuels today.

1 What is a biofuel? ⑦

B Biofuels include waste material such as wheat stalks, peanut shells or branches left over when trees are cut down for wood. These can be burnt in power stations such as this one in Scotland.

C British Sugar's bioethanol fuel plant in Norfolk uses sugar beet.

Chemists can manufacture biofuels from raw materials. **Ethanol** is made by processing wheat, sugar cane or sugar beet. Ethanol can be mixed with petrol for use as a fuel in car engines. Using ethanol helps to reduce the demand for petrol and so conserves crude oil supplies.

Biodiesel is a fuel made from vegetable oils by chemical reactions. The oil can be produced from oil seed rape or soya beans or used cooking oil from restaurants. Diesel engines can run on biodiesel or a mixture of biodiesel and normal diesel oil.

Advantages and disadvantages

Biofuels are a **renewable** resource. For example, so long as trees are not removed from a forest faster than more can be grown, our supply of wood will never run out. Crops grown to make into ethanol or biodiesel can be grown continuously.

Biofuels may also help to reduce the overall amount of carbon dioxide that human activity puts into the atmosphere. Plants take in carbon dioxide from the air when they photosynthesise. When plants are made into biofuel and burnt, this carbon dioxide is returned to the atmosphere by combustion. So in theory a biofuel can be **carbon neutral** (overall, it does not add carbon dioxide to the atmosphere). However, energy is needed to make fertiliser for the crops, to harvest them, to process them into biofuel and to transport the biofuel to where it is needed. The energy for most of these processes is obtained from fossil fuels, so biofuels may not be carbon neutral when the manufacturing and distribution processes are considered.

D *Some modern cars can run on 85% ethanol mixed with only 15% petrol.*

Biofuels could replace much of the fossil fuel that we use. This would require a lot of crops to be grown specially to make the biofuel. However, if we grow a lot of crops to make biofuel this will mean there is less farmland for growing food or it may lead to forests being cleared to plant biofuel crops instead.

2 How can ethanol reduce the demand for petrol?

ResultsPlus
Watch Out!

Biofuels release carbon dioxide when they burn, but they only release the amount of carbon dioxide taken in by the plants used to produce them. They can be thought of as carbon neutral.

Skills spotlight

What factors do you think should be taken into account before a decision is taken to clear some rainforest to plant palm oil trees for biofuel?

3 What can biodiesel be made from?

4 Which biofuels could be used in vehicles?

5 a Write down two advantages of biofuels compared with fossil fuels.
b Write down two disadvantages compared with fossil fuels.
c Do you think we should use more biofuels in the UK? Explain your opinion.

6 Explain why most biofuels may not be 'carbon neutral'.

Learning Outcomes

5.17 Describe biofuels as being possible alternatives to fossil fuels

5.18 Recall that one example of a biofuel is ethanol obtained by processing sugar cane or sugar beet and that it can be used to reduce the demand for petrol

5.19 Evaluate the advantages and disadvantages of replacing fossil fuels with biofuels, including: **a** the fact that biofuels are renewable **b** that growing the crops to make biofuels requires land and may affect the availability of land for growing food **c** the balance between the carbon dioxide removed from the atmosphere as these crops grow and the carbon dioxide produced when they are transported and burned

HSW **13** Describe the social, economic and environmental effects of decisions about the uses of science and technology

>>>>>>>>>>>>>>>>>>>>>>>>>>>>>>

What fuel is used in rockets?

C1.29 Choosing fuels

How do we choose the best fuel to use?

This is the *Buckeye Bullet 2*. It holds the current speed record of 508 km/h for a vehicle powered by a fuel cell. It was built by students from Ohio State University.

A *The fastest car powered by a fuel cell*

Fuels and pollution

In a rocket, hydrogen and oxygen burn to release energy. Hydrogen and oxygen can also react and release energy without burning in a device called a **fuel cell**. A hydrogen fuel cell produces electricity to run an electric motor. Water is the only waste product. Since it only produces water when used as a fuel, hydrogen is often called a 'clean' fuel.

B *Most rockets use hydrogen as a fuel – they also need to carry liquid oxygen with them to allow the hydrogen to burn.*

Most of the fuels we use cause some pollution. Natural gas (which is mainly methane), petrol, kerosene and diesel oil (obtained from crude oil) are all non-renewable fossil fuels. They all produce carbon dioxide and water when they burn completely. If incomplete combustion occurs, any of these fuels can produce carbon monoxide and soot. Some fuels contain impurities such as sulfur. Sulfur forms sulfur dioxide when it burns.

Biofuels are renewable fuels. They burn to produce carbon dioxide and water, but they can also produce carbon monoxide and soot. Biofuels such as wood also leave a lot of ash when they burn.

What makes a good fuel?

Pollution is not the only factor to consider when choosing a fuel. Other factors are:
- how easily it burns
- how much heat energy it produces (Figure C shows how much energy is released by different fuels)
- how easy it is to store and transport.

A good fuel must burn easily. It is much easier to light a cooker that uses natural gas than it is to light a coal fire. However, fuels that are easy to light can also be more dangerous if they are not stored and transported safely. Leaking natural gas can cause explosions.

1 Describe two ways in which hydrogen can be used as a fuel in engines.

2 Why does a fuel cell produce less pollution than a car engine running on petrol or diesel?

3 What effects do these gases have on the environment:
a carbon dioxide
b sulfur dioxide?

Fuel	Energy released (MJ/kg)
hydrogen	141.8
methane	55.5
petrol	47.3
kerosene	46.2
diesel	44.8
coal	27.0
wood	15.0

C *Energy released by different fuels. 1 megajoule (MJ) = 1 000 000 J*

Coal and other solid fuels are easy to store but have to be transported by lorry or train. Liquid and gas fuels are stored in tanks and can be transported in tankers or sent through pipes. Gas fuels such as methane and hydrogen have to be stored at high pressure to reduce the sizes of the tanks needed to store them.

Hydrogen or petrol?

Buses with hydrogen-powered fuel cells are used in some cities.

D Steam trains used coal as a fuel – the coal had to be shovelled into the furnace by hand. Today many trains use diesel.

However, the hydrogen used in these fuel cells is mainly produced from natural gas, and this process releases carbon dioxide.

Before cars with fuel cells can become widely used, hydrogen has to be easily and economically available. Filling stations would have to be converted to store and sell compressed hydrogen as well as petrol and diesel. This will cost a lot of money. People will not start to use fuel-cell powered cars until it is easy to buy hydrogen, and energy companies do not want to start converting all their filling stations until there are plenty of customers!

4 Give the names of two fuels that must be stored at high pressure.

Skills spotlight

Suggest how governments could persuade more people to buy cars powered by fuel cells. Explain whether you think using fuel cells instead of normal engines is a good idea.

Results Plus
Watch Out!

In a fuel cell, hydrogen and oxygen are combined *without burning* to produce electricity.

5 Give two reasons why coal is no longer used as fuel for trains.

6 Suggest why hydrogen, rather than petrol or kerosene, is used as the fuel for rockets.

7 Construct an argument for or against using hydrogen instead of petrol in cars. Remember to include advantages and disadvantages.

8 Explain why some people would like petrol to be replaced with hydrogen for use in fuel cells for car engines, and why this may take some time to happen.

Learning Outcomes

5.20 Demonstrate an understanding of the factors that make a good fuel, including: **a** how easily it burns **b** the amount of ash or smoke it produces **c** the comparative amount of heat energy it produces (calculations involving conversion to joules are not required) **d** how easy it is to store and transport

5.21 Recall that a simple fuel cell combines hydrogen and oxygen to form water and that this reaction releases energy

5.22 Evaluate the advantages and disadvantages of using hydrogen, rather than petrol, as a fuel in cars

5.23 Describe petrol, kerosene and diesel oil as non-renewable fossil fuels obtained from crude oil and methane as a non-renewable fossil fuel found in natural gas

HSW **13** Explain how and why decisions that raise ethical issues about uses of science and technology are made

C1.30 Investigating fuels

Do all fuels release the same amount of energy?

The first hot air balloon to carry people was tethered to the ground and used air heated by a fire on the ground beneath it. A month later, the first balloon flight carried enough fuel for the balloon to fly nearly 10 km. The balloon had to land because sparks from the fire were beginning to set light to the balloon itself!

B *Most modern hot air balloons use hydrocarbon fuels.*

Modern hot air balloons carry cylinders of liquid propane, a hydrocarbon fuel. Burning the propane produces the hot air that keeps the balloon airborne. In 1999 a hot air balloon flew around the world without refuelling. A fuel that provides a lot of heat energy for each gram of fuel is very important to allow balloons to fly long distances.

Different hydrocarbon fuels have different numbers of carbon atoms in each molecule. Different hydrocarbons release different amounts of energy when they burn.

A *The first untethered hot air balloon to carry people flew 1783. The balloon was tethered to the ground for the fi flight, and used hot air from a fire on the ground benea it, and an on-flight brazier kept the balloon in the air for 25 minutes.*

Your task

You are going to plan an investigation that will allow you to find out how the amount of heat given out by a fuel changes depending on how many carbon atoms are in each molecule of the fuel. You are going to test the hypothesis that when you heat water, the temperature rise you get for each gram of a fuel depends on the number of carbon atoms in the molecules of the fuel. Your teacher will provide you with some materials to help you organise this task.

Learning Outcomes

5.24 Compare the temperature rise produced when the same volume of water is heated by different fuels

Build Better Answers

When completing an investigation like this, one of the skills you will be assessed on is your ability to *evaluate your method*. There are 6 marks available for this skill. Here is one student extract that focuses on this skill. Other skills that you need for the assessment are dealt with in other lessons.

| Student extract 1 | A good response for this skill |

> I think that one of the main weaknesses of the experiment was that not all the heat from the burner ended up heating the water. One way to solve this was to make sure there were no drafts around the burner or use a heat shield to make sure that as little heat was lost as possible. Some fuels took longer to heat the water than others and this meant that proportionally more of the heat from these fuels was lost. This would make it look like these fuels gave out a lot less energy than they did. One way to solve this would be to heat the water for a smaller temperature rise. This would mean that the fuels would not be burning for as long and there would be less heat lost to the environment. This would improve the quality of my data. The hypothesis is that when you heat water, the temperature rise you get for each gram of a fuel depends on the number of carbon atoms in the molecules of the fuel. If we use a heat shield and make sure that as much of the heat as possible is used for heating the water then we can make sure that we are getting a better picture of what energy is given out from each fuel. Another problem was that the fuel evaporated from the burner quite quickly. To solve this we should remember to put the cap onto the burner when we are not using it.

This is a clear example of identifying a problem and then suggesting a sensible way to solve it.

You need to explain how the weaknesses in the method affects the quality of the data that you have gathered.

As well as giving a way to solve the problem with the method you need to explain how this improves the data that you are gathering.

Explain how your improvements will help you to prove or disprove your hypothesis.

Give more than one example of a weakness and an improvement.

To access 4 marks

- Describe strengths or weaknesses in your method
- Provide reasons for any anomalies
- Suggest how to improve your method and explain how this will improve the quality of the evidence you could collect

To access 6 marks

You also need to:

- Relate your comments back to the original hypothesis of the investigation

What are alkanes and alkenes?

This house was destroyed by a natural gas explosion. A strong-smelling substance is added to natural gas as a safety precaution to help people smell accidental gas leaks before they cause explosions.

A *This house was destroyed by a gas explosion.*

Alkanes

When natural gas is extracted, it contains hydrocarbon compounds called methane, ethane and propane – the gas is processed so that the gas piped to homes is almost all methane. These compounds are all part of a group of molecules called **alkanes** which are mainly used as fuels. Other alkanes are found in crude oil.

A carbon atom can form **bonds** with a maximum of four other atoms. Each carbon atom in an alkane molecule is bonded to four other atoms with single bonds. Hydrocarbons with carbon–carbon bonds that are all single bonds are called **saturated** hydrocarbons. Alkanes are saturated hydrocarbons.

Methane is the simplest alkane, with only one carbon atom joined to four hydrogen atoms in its molecule. Ethane has two carbon atoms in each molecule, and propane has three carbon atoms per molecule. You can see from Figure B that each carbon atom is joined to four other atoms by single bonds.

> **?**
>
> 1 a What is an alkane?
> b Name three alkanes, and write down their molecular formulae.
>
> 2 Draw the structure of a molecule of ethane.
>
> 3 Why are alkanes called:
> a hydrocarbons
> b saturated hydrocarbons?

> **ResultsPlus**
> **Watch Out!**
>
> Alkanes are saturated by hydrocarbons with single C–C bonds. Alkenes are unsaturated hydrocarbons with one double C=C bond.

methane CH_4

ethane C_2H_6

propane C_3H_8

B *Methane, ethane and propane are examples of saturated hydrocarbons called alkanes.*

Alkenes

Some hydrocarbon molecules have a **double bond** between two of the carbon atoms. Hydrocarbons with double bonds in their molecules are **unsaturated** hydrocarbons. The carbon–carbon double bond is not 'saturated' because the carbons are not bonded to the maximum number of other atoms.

Alkenes are hydrocarbons with one double bond between carbon atoms. Alkenes are unsaturated hydrocarbons.

C *Ethene and propene are examples of unsaturated hydrocarbons called alkenes.*

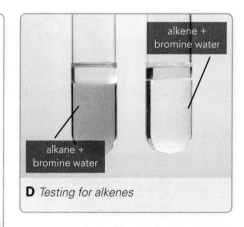

D *Testing for alkenes*

Alkane or alkene?

The **bromine test** is used to find out if a compound contains double bonds. **Bromine water** is bromine dissolved in water and is usually orange (although it can be pale yellow to red-brown depending on the amount of bromine dissolved). If it is mixed with a saturated hydrocarbon, there is no colour change. However, if it is mixed with an unsaturated hydrocarbon, it reacts with the alkene and decolorises (the colour fades away).

ethene   +   bromine water   →   colourless liquid
(colourless)   (orange)

ethane   +   bromine water   →   orange-coloured liquid
(colourless)   (orange)

4 What is the difference between an alkane and an alkene?

5 Why are alkenes called unsaturated hydrocarbons?

6 a Name two alkenes, and write down their formulae.
  b Draw the structures of the two alkenes you have named.

7 Explain how you could use bromine water to find out which of two liquids is an alkane and which is an alkene.

Skills spotlight

There is an internationally agreed way of naming substances. The names also help to describe which elements are present in the molecule and how the atoms are arranged.

a How do the names of the alkanes and alkenes tell you how many carbon atoms they have in a molecule and whether or not they have any double bonds?

b Why is it important to have an internationally agreed way of naming compounds?

Learning Outcomes

5.25 Recall that alkanes are saturated hydrocarbons, which are present in crude oil

5.26 Recall the formulae of the alkanes methane, ethane and propane, and draw the structures of these molecules to show how the atoms are bonded together (no further knowledge of bonding is required in this unit)

5.27 Recall that alkenes are unsaturated hydrocarbons

5.28 Recall the formulae of the alkenes ethene and propene and draw the structures of their molecules to show how the atoms are bonded together (no further knowledge of bonding is required in this unit)

5.29 Describe how bromine water is used to distinguish between alkanes and alkenes

HSW 11 Present information using scientific conventions and symbols

How do oil companies make sure we have enough petrol?

How can other compounds be made from crude oil?

We use more petrol than can be obtained directly by fractional distillation of crude oil. The crude oil fraction that is used as petrol makes up only about 15% of the crude oil. So why don't we have a permanent shortage of petrol?

A *Fuel shortages cause long queues at filling stations*

Crude oil is separated into different fractions to produce useful fuels and other products. However, the longer hydrocarbons in some fractions can be split into shorter hydrocarbons, which are more useful. This is done by a process called **cracking**.

In cracking, long-chain alkanes are heated and break up into smaller molecules. Cracking is an example of a thermal decomposition reaction, because molecules are being broken up by heat. The energy for heating the alkanes comes from burning fossil fuels.

longer chain alkanes → shorter chain alkanes + alkenes

B *Cracking hydrocarbons*

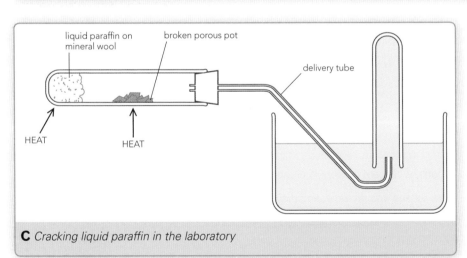

C *Cracking liquid paraffin in the laboratory*

Cracking long-chain alkane molecules produces shorter-chain alkanes and alkenes. If you look at Figure B carefully, you will see that there are exactly the same numbers of carbon and hydrogen atoms before and after cracking. However, because the carbon atoms are now in two separate molecules, there are not enough hydrogen atoms to go around. This is why one molecule is unsaturated (the carbon atoms have a double bond between them).

Cracking produces shorter-chain alkanes that are used as fuels (like petrol) or used for making other substances. Most of the alkenes produced are used for making plastics, but some ethene is also used for ripening fruit.

Why is cracking needed?

Crude oil obtained from different oil fields contains different mixtures of hydrocarbon molecules. Figure C shows the percentage of each fraction in a typical sample of crude oil. It also shows the demand from customers for each fraction. If oil companies just sold the fractions as they come out of a fractionating column, they would have too much of fractions such as kerosene and fuel oil. To make supply match demand, the oil companies crack the longer molecules.

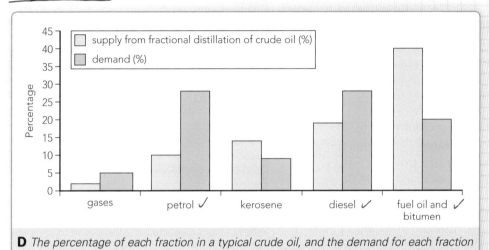

D *The percentage of each fraction in a typical crude oil, and the demand for each fraction*

1 What is cracking?

2 Why do oil companies crack long-chain alkanes?

H 3 What is formed when a large alkane molecule is cracked?

4 Give two uses for the products of cracking.

5 Look at Figure D. For which fractions is: **a** supply greater than demand **b** demand greater than supply?

H 6 Suggest why the demand for petrol and diesel is higher than the demand for other fractions of crude oil.

H 7 Write a balanced equation for the cracking reaction shown in Figure B.

8 You have two liquids – one is a fraction of crude oil and one is a fraction that has been cracked. Explain how you could find out which one is which.

Results Plus
Watch Out!

If oil companies did not use cracking to make more petrol, they would have to extract more oil from the ground. This would be more expensive and lead to a lot of wasted bitumen and fuel oil.

Skills spotlight

Scientists often use models to explain complex ideas. Make a model of a long alkane molecule using paper clips or a modelling kit. Then break the model to show what happens during cracking.

a Write some notes to help you explain to younger students what happens in cracking.

b What are the advantages and disadvantages of your model? How well does it represent the molecules?

Learning Outcomes

5.30 Describe how cracking involves the breaking down of larger saturated hydrocarbon molecules (alkanes) into smaller, more useful ones, some of which are unsaturated (alkenes)

H **5.31** Explain why cracking is necessary, including by using data on the composition of different crude oils and the demand for fractions in crude oil

5.32 Describe the cracking of liquid paraffin in the laboratory

HSW **3** Describe how phenomena are explained using scientific models

What are polymers and how are they made?

Polythene helped the Allies to win the Second World War. Polythene was developed in the UK just before the war and was a better and much lighter electrical insulator for wires and cables than the materials in use at the time. This helped to make the newly invented radar-sets light enough to be fitted into RAF fighter aircraft so that pilots could detect the enemy at night.

A A World War II fighter ready for take-off

> **1** What is a:
> **a** monomer
> **b** polymer?

The correct name for polythene is **poly(ethene)**. Poly(ethene) is an example of a **polymer** – these are substances made up of thousands of relatively simple repeating units. **Monomers** are substances whose molecules react together to form polymers. Poly(ethene) is made from lots of ethene monomers.

Results Plus
Watch Out!

You need to be able to recognise or draw the structure of the monomer molecules and formulae for ethene (C_2H_4) and propene (C_3H_6).

Some polymers occur naturally, such as proteins and cellulose (the substance found in plant cell walls). Other polymers are manufactured from the products of cracking crude oil. Manufactured polymers are often called **plastics**. Poly(ethene) is used to make plastic bags and plastic bottles.

When **polymerisation** reactions occur, the monomer molecules react together to form long-chain molecules. A poly(ethene) molecule can have thousands, or even millions, of carbon atoms in it.

B How poly(ethene) is made

Skills spotlight

H Look at Figure C. Describe how this equation is different to other balanced equations you have seen. Explain why the convention used is useful.

Because the number of ethene molecules (n) that join to make one molecule of poly(ethene) is very large, we write the equation as shown in Figure C. The repeating unit is shown in brackets with the subscript n.

C Many (n) ethene monomers polymerise (react together) to form a long chain of poly(ethene) with n repeat units.

Poly(ethene) is made from ethene. Other polymers are made from other unsaturated molecules. Figure D shows the monomers that make **poly(propene)**, **poly(chloroethene)** and **poly(tetrafluoroethene)**. These polymers all have different uses, which are related to their properties. New materials with new uses are being developed all the time.

propene monomer
[forms poly(propene)]

chloroethane monomer
[forms poly(chloroethene)]

tetrafluoroethene monomer
[forms poly(tetrafluoroethene)]

D *The monomers that form three common polymers*

Polymer (and common name)	Properties	Uses
poly(ethene) polythene or polyethylene	flexible, cheap, good insulator	plastic bags, plastic bottles, cling film, insulation for electrical wires
poly(propene) polypropene or polypropylene	flexible, shatterproof, high softening point	buckets and bowls
poly(chloroethene) **PVC**	tough, cheap, long-lasting, good insulator	window frames, gutters, pipes, insulation for electrical wires
poly(tetrafluoroethene) **PTFE** or **Teflon**®	tough, slippery, resistant to corrosion, good insulator	non-stick coatings for saucepans, bearings and skis, containers for corrosive substances, stain-proofing carpets, insulation for electrical wires

E *Some polymers and their uses*

2 What polymer is made from the monomers: a ethene b chloroethene?

3 Why are many plastic bags made from poly(ethene)?

4 Which polymer in Figure E would you use to make: a a coating for a table that stops dust sticking to it b a seaside bucket and spade c the frame for a small greenhouse?

5 Explain the reasons for your choices in question 4.

6 Use Figure D to help you to write equations (similar to the one in Figure C) for making these polymers from their monomers: a poly(propene) b poly(chloroethene).

7 Describe the polymerisation of poly(ethene). Include a drawing for the structure of the monomer molecule.

Learning Outcomes

5.33 Recall that: *a* many ethene molecules can combine together in a polymerisation reaction *b* the polymer formed is called poly(ethene) (conditions and mechanisms not required but Ⓗ equations are required)

5.34 Describe how other polymers can be made by combining together other monomer molecules, to include poly(propene), poly(chloroethene) (PVC) and PTFE

5.35 Relate uses of the polymers poly(ethene), poly(propene), poly(chloroethene) (PVC) and PTFE to the properties of the compounds

HSW 11 Presenting information using scientific conventions and symbols

Why do people make clothes out of materials from waste plastic bottles?

Why should we try to recycle polymers?

This fleece is made from recycled polymers. Making clothes like this helps to reduce the amount of polymer waste that gets sent to landfill sites. It also reduces the amount of raw materials needed to make new fabrics.

A Clothing made from recycled polymers.

1 What does biodegradable mean?

Materials such as wood and paper are **biodegradable**. This means that they rot because microbes can feed on them. Most manufactured polymers are useful for many purposes because they are *not* biodegradable and so they last for a long time. However, this also means that they do not rot when they are thrown away.

Most of the rubbish in the UK goes to landfill sites. The non-biodegradable plastic materials in the rubbish will last for many years. Some waste is **incinerated** (burned), and the energy released can be used to generate electricity. However, many plastics produce toxic substances when they burn. Most of these toxic substances can be removed from the waste gases, but this forms toxic ash, which then must be disposed of safely.

B The plastic waste in this landfill site will not rot away.

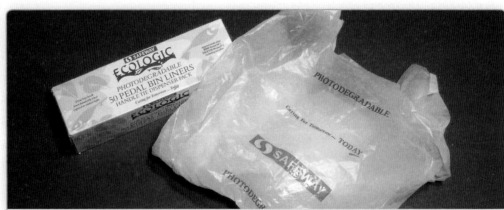

C Chemists have developed some new polymers that will rot away.

2 Most manufactured polymers are not biodegradable. Why is this:
a an advantage
b a disadvantage?

3 Waste can be incinerated. Why can incinerating polymers cause problems?

Chemists are also developing polymers that will biodegrade, such as the ones shown in Figure C. If more plastic materials are made of these polymers, they will rot within a few years if they end up in a landfill site. However, this takes time, so it would be better if we reduce the amount of plastic sent to landfill sites as much as possible.

We can reduce the amount of waste that goes to incinerators or landfill sites by reducing the amounts of the materials we use and also by reusing materials. For example, we can reuse plastic bags rather than just throwing them away after a single use. If an item cannot be reused any more, the material it is made of can be recycled. This means that the materials in it are processed and used to make new objects. Reusing and recycling both mean that raw materials (like crude oil) should last longer.

A lot of paper, glass and metal waste is already recycled in the UK. It is more difficult to recycle polymers because the waste needs to be sorted into different types of polymer before each type can be made into new objects. Plastic bottles and packaging materials are marked with a symbol to show the type of plastic.

ResultsPlus
Watch Out!

Reuse means using an item or material again, not throwing it away after a single use. Recycle means the materials in an item are processed and used to make new objects!

symbol	polymer	uses
1 PET	poly(ethylene) terephthalate	some bottles, food trays, duvet fillings
2 HDPE	high-density poly(ethene)	some bottles, buckets
3 PVC	poly(chlorothene)	soft toys, window frames
4 LDPE	low-density polythene	cling film, bags
5 PP	poly(propene)	crisp packets, carpet, ropes
6 PS	poly(styrene)	egg boxes, foam packaging
7 OTHER	other polymers	

D Polymer recycling symbols

E The recycling symbols on polymers are not always very easy to see.

4 a Explain the difference between 'reusing 'and 'recycling'.
b Why is it better to reuse something than to recycle it? Explain in as much detail as you can.
c Why is it more difficult to recycle plastic than glass?

5 Milk used to be delivered to every home in glass bottles, and the empty glass bottles were collected to be reused. Today, most people buy milk in plastic bottles from supermarkets. What are the advantages and disadvantages of each of these systems?

Skills spotlight

What social, economic and environmental factors should a local council consider when deciding whether or not to build an incinerator or whether to start collecting plastics for recycling?

Learning Outcomes

5.36 Recall that most polymers are not biodegradable, persist in landfill sites and that many produce toxic products when burnt

5.37 Explain how some problems associated with the disposal of polymers can be overcome: a by recycling b by developing biodegradable polymers

HSW 13 Describe the social, economic and environmental effects of decisions about the uses of science and technology

These questions are indicative of the type of questions used in the exam. Refer to page 6 for information on the grades.

The Earth's atmosphere

1. The Earth's atmosphere is made up of several gases.

   (a) Draw a straight line to join the name of the gas with the percentage of it present in the atmosphere.

   carbon dioxide •

   nitrogen •

   oxygen •

   • 0.04%

   • 0.9%

   • 21%

   • 78%

   (2)

   (b) In the early atmosphere, there was much more water vapour and carbon dioxide than now.

   (i) Where did the gases that made up the early atmosphere come from?

   A the sun

   B the oceans

   C growth of plants

   D volcanic activity

   (1)

   (ii) Other than water vapour and carbon dioxide, name another gas that was thought to be present in the early atmosphere.

   A chlorine

   B argon

   C methane

   D oxygen

   (1)

   (c) Over time, the Earth's atmosphere cooled down. Explain what happened to the water vapour in the Earth's atmosphere as the Earth cooled down.

   (2)

   (d) About 1.7 billion years ago, oxygen started to appear in the atmosphere. Explain why oxygen levels have increased since then to today's level.

   (2)

Natural forms of calcium carbonate

2. Limestone is a natural form of calcium carbonate. Heating limestone starts a reaction which makes calcium oxide and carbon dioxide.

   (a) Complete the word equation for this reaction.

   calcium carbonate → _____ + _____

   (1)

   (b) From the list, select the term used to describe this type of reaction.

   A exothermic reaction

   B oxidation

   C reduction

   D thermal decomposition

   (1)

(c) Match up the following substances with their chemical formula. Each formula may be used once, more than once, or not at all.

calcium carbonate •

calcium oxide •

carbon dioxide •

• CO_2
• $CaCO_3$
• CaO
• $Ca(OH)_2$

(2)

(d) The photograph shows a limestone quarry.

Give two problems caused by quarrying limestone. (2)

(e) Limestone is a sedimentary rock. The photograph shows the limestone cliffs in Bridport, Dorset.

Describe, using evidence from the photograph, how sedimentary rocks are formed. (3)

Extracting and using copper

3. In the past, copper could be found as the free metal, but today much of the copper we use is extracted from ores such as malachite.

   (a) What is an ore? (1)

   (b) In terms of reactivity, why can copper sometimes be found as the free metal? (1)

   (c) In the first stage of extracting copper, malachite (copper carbonate) is converted into copper oxide. The following word equation shows what happens:

   copper carbonate → copper oxide + carbon dioxide

   What has to be done to copper carbonate to bring about this change? (1)

   (d) The word equation for the second stage is:

   copper oxide + carbon → copper + carbon dioxide

   The oxygen is removed from copper oxide. What name is given to this process? (1)

   (e) Match some uses of copper with some of its properties. Each property may be used once, more than once or not at all. (2)

use of copper	property of copper
	• ductile
coins •	
	• good conductor of electricity
electrical wire •	
	• good conductor of heat
saucepans •	
	• low in reactivity

   (f) These days, most of the copper that is thrown away is recycled. Explain the environmental and economic benefits of recycling copper. (3)

Acids and salts

4. Acids react with bases to form salts. The acid found in our stomachs is mostly hydrochloric acid.

   (a) Draw straight lines to match each acid up with its formula and with the type of salt it forms.   (2)

formula	acid	salt
HCl •	• nitric acid	• sulfate
HNO$_3$ •	• hydrochloric acid	• chloride
H$_2$SO$_4$ •	• sulfuric acid	• nitrate

   (b) Hydrochloric acid can be electrolysed using the apparatus shown in the diagram.

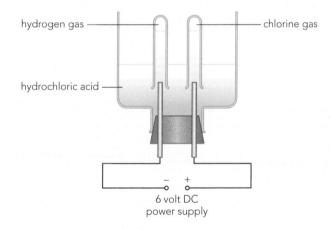

hydrogen gas ———————————— chlorine gas

hydrochloric acid —

− +

6 volt DC
power supply

   During electrolysis, the electrical current decomposes the hydrochloric acid into hydrogen gas and chlorine gas.

   (i) Describe the test for hydrogen gas.   (2)

   (ii) Describe the test for chlorine gas.   (2)

   (c) Indigestion is caused by excess acid in the stomach. People take antacid tablets to treat the indigestion. The antacid tablet reacts with the stomach acid and neutralises it. A student was trying to find out if chewing an antacid tablet made the reaction faster.

   Explain, in detail, how you could safely carry out the investigation to find out if crushing the tablet into a powder makes the reaction faster.   (6)

Hydrocarbon fuels

5. Many hydrocarbons are used as fuels.

    (a) Below are the structures of molecules of two hydrocarbons.

    hexane   octane

    Name the two elements that are combined together in molecules of alkanes. (2)

    (b) Hexane is a liquid. During complete combustion, oxygen reacts with hexane to form carbon dioxide and water.

    (i) Write the word equation for the reaction. (1)

    (ii) What different product would form during the combustion of hexane if there was a restricted amount of oxygen available? (1)

    (iii) State two factors that make a good fuel. (2)

    (c) The amount of energy released when a fuel burns can be determined using the apparatus shown in the diagram.

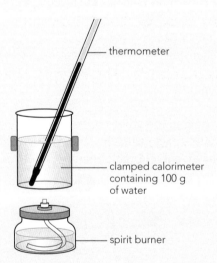

thermometer

clamped calorimeter containing 100 g of water

spirit burner

    You have been given two hydrocarbon fuels to investigate, hexane and octane. Explain, in detail, how you could use the apparatus safely to see which fuel gives out the most heat energy. (6)

Rocks

1. The photographs show samples of a sedimentary rock, a metamorphic rock and an igneous rock.

Rock P   Rock Q   Rock R

(a) Rock P is sedimentary. What evidence could you **see** when examining a rock to suggest that
it is sedimentary?   (1)

(b) Rock Q is metamorphic. Metamorphic rocks are formed when:

   A   magma cools
   B   layers of small rock fragments build up
   C   existing rocks are changed by heat and pressure
   D   sedimentary rocks are melted and solidified   (1)

(c) Rock R is igneous. It contains crystals. Explain how magma forms igneous rocks with different
sized crystals.   (2)

(d) Limestone is a sedimentary rock composed of calcium carbonate, $CaCO_3$. Calcium carbonate
forms calcium oxide, CaO, and a gas when heated strongly.

   (i) Complete the balanced equation to show the thermal decomposition of calcium carbonate. (2)

   $CaCO_3 \rightarrow$ _____

   (ii) The United Kingdom's limestone supplies come from quarries. Give an advantage and a
   disadvantage of living close to a limestone quarry.   (2)

Hydrochloric acid

2. Various gases are released during electrolysis, including chlorine and hydrogen. The diagram shows the apparatus that can be used to electrolyse dilute hydrochloric acid, HCl.

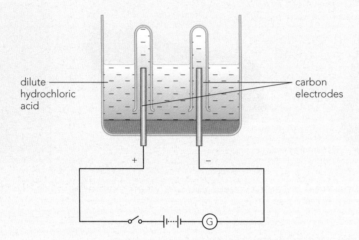

(a) (i) Chlorine is used to sterilise drinking water. What property of chlorine makes it suitable for sterilising water?

      A it is a halogen
      B it is a gas
      C it is toxic
      D it is reactive   (1) ▮

   (ii) Describe the test for chlorine.   (2) ▮

(b) What is produced at the cathode and at the anode in the electrolysis of dilute hydrochloric acid?

	Cathode	Anode
A	chlorine	oxygen
B	chlorine	hydrogen
C	hydrogen	chlorine
D	oxygen	hydrogen

    (1) ▮

(c) Hydrochloric acid reacts with copper carbonate, $CuCO_3$. The salt formed is copper chloride, $CuCl_2$. Write a balanced equation for this reaction.   (2) ▮

(d) You are given three different indigestion tablets, A, B and C. Describe how you could test the three tablets to see which was the most effective at neutralising acid.   (3) ▮

Metals and alloys

3. This information is about shape memory alloys.

A shape memory alloy 'remembers' its original shape. It can be deformed by applying heat, but will then return to its original shape.

Shape memory alloys have applications in the medical and aerospace industries. Titanium-containing shape memory alloys are used to make frames for spectacles.

(a) (i) What is an alloy? (1)

(ii) Nitinol is a shape memory alloy. It contains titanium. Name the other element in nitinol. (1)

(iii) Spectacle frames made from alloys are strong. Give another reason that shape memory alloys are used to make spectacle frames. (1)

The table shows three metals in order of reactivity.

Most reactive	aluminium
	iron
Least reactive	gold

(b) (i) Explain why aluminium must be extracted by electrolysis, even though this method is expensive. (2)

(ii) State why an aluminium alloy, rather than steel, is used to make aircraft bodies. (1)

(iii) Evaluate the advantages and disadvantages of extracting iron from iron ore compared to recycling scrap iron. (4)

Fuels

4. Fuels are substances that burn to release heat energy. The photograph shows a gas fire. Gas fires often burn methane gas to produce carbon dioxide and water. Sometimes a gas fire may produce carbon monoxide.

(a) (i) State why carbon monoxide is sometimes formed in fires rather than carbon dioxide. (1) ▇

   (ii) Why is the formation of carbon monoxide a safety hazard? (1) ▇

(b) (i) Propane can be used as a fuel. Draw one molecule of propane, showing all covalent bonds. (2) ▇

   (ii) Explain why propene can form a polymer but propane cannot. (2) ▇

(c) The amount of heat energy released when different fuels burn can be measured. You are given 10 cm³ samples of the liquid fuels ethanol, propanol, paraffin and octane. Explain how you could carry out an experiment to find out which 10 cm³ sample of fuel gives off the most heat energy. (6) ▇

Evolution of the Earth's atmosphere

5. The Earth's early atmosphere is thought to have contained little or no oxygen, a large amount of carbon dioxide and water vapour, and small amounts of other gases.

    (a) The pie chart shows the composition of dry air in **today's** atmosphere. Complete the table to show the percentage of the three main gases in dry air.

(2)

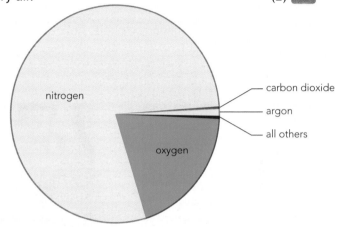

Gas	Percentage in dry air (%)
nitrogen	
oxygen	21
argon	

(b) The apparatus shown below can be used to find the percentage of oxygen in a sample of air. The air is passed backwards and forwards over the heated copper. When no further change occurs, the apparatus is allowed to cool back to the original temperature.

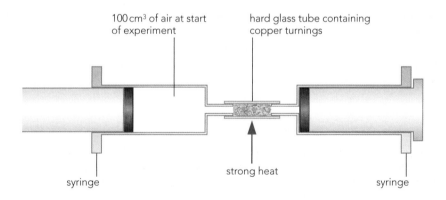

During one experiment the copper turns black, and the volume of gas reduces from 100 cm^3 to 80 cm^3 at the end of the experiment.

(i) Explain why the copper turns black.

(2)

(ii) Explain why the volume of gas reduces from 100 cm^3 to 80 cm^3.

(2)

(c) The composition of the atmosphere today is different from the early atmosphere. Describe how the proportion of oxygen, carbon dioxide and water vapour has changed from the early atmosphere to the present day. Suggest why scientists might be uncertain about this information.

(6)

Here are three student answers to the following question. Read the answers together with the examiner comments around and after them.

Rocks can be sedimentary, metamorphic or igneous. The photographs below show examples of these three types of rock.

Describe, in as much detail as possible, how the rocks were formed and how they differ from each other in appearance. **(6)**

limestone
(sedimentary)

slate
(metamorphic)

granite
(igneous)

Student answer 1   **Extract typical of a level ① answer**

Be sure to spell scientific terms correctly – 'granit' and 'cristals' are both mis-spelled.

> Limestone is crumbly and soft, and the granit and slate are harder.
>
> Granit is made of cristals and is used to make kitchen worktops.
>
> Limestone is made by cooling rock down and granit is made from sediments.

This shows some knowledge of the description of the three categories of rocks, but you can't use the word 'harder' to describe what something looks like.

It is important to remember facts correctly. This is not how the rocks were formed.

Examiner summary
This answer includes some correct details about what the rocks are like, but not about how the rocks were formed. Be very careful when spelling any scientific terms.

There are some appropriate descriptions here but they could go further.

> Granite is hard and made up of crystals. Slate is made up of layers. These layers can be split. Limestone is a hard rock, but it is softer than the other two. Limestone is made up of sediments falling on top of each other and is pushed together. It contains fossils.

This is the correct description of how limestone was formed but could use more scientific words in the explanation.

Examiner summary

The details in this answer are good, but it could include more to improve the answer further. If some of the short sentences were joined up they would be easier to read.

This shows good knowledge of how sedimentary rocks were formed.

> Limestone is a hard, crumbly rock made from sediments that have been crushed together. It was formed at the bottom of the sea and sometimes it contains fossils. Granite is a hard rock and contains crystals. These crystals are locked together. Granite is formed by cooling down molten rock. Slate is a metamorphic rock which is formed by heating or pressing together the other rocks. Slate is hard and made up of layers.

Good description of how igneous rocks are formed.

Good description of how metamorphic rocks are formed.

Examiner summary

This is an excellent answer at this level. It includes details about how the rocks are formed and about what they are like. It is clear and easy to read, with good spelling, punctuation and grammar and correct technical language.

 Results**Plus**

Move from level ① to level ②

Try not to fill the space with things that do not answer the question. You may find it easier to follow on one aspect of the question. Here, this may be the appearance of the rocks – are they in layers, in crystals?

Move from level ② to level ③

Make sure you answer all aspects of the question – here this means giving some detail on how each rock formed. An excellent answer is one which has not only the details but also is given in complete sentences using the correct technical language.

Here are three student answers to the following question. Read the answers together with the examiner comments around and after them.

Question | **Combustion of methane** | Grade | G–D

The fuel used in some gas fires is methane.

The complete combustion of methane produces carbon dioxide and water, as well as heat energy. In faulty gas fires, incomplete combustion of methane can take place.

Explain the problems that can be caused by the complete and the incomplete combustion of methane in a gas fire. (6)

Student answer 1 | Extract typical of a level ① answer

This gives some information about one of the products of incomplete combustion.

> The gas that is formed is called Carbon monoxide. This is a dangerous gas. It's a killer. It gets formed cos the gas fire is fawlty.

Be sure to use correct spelling (e.g. 'faulty', not 'fawlty') and don't use shorthand, e.g. 'cos'.

This highlights one problem caused by the gas.

Examiner summary
This answer includes some knowledge about incomplete combustion. To improve this answer, explain how these the products formed and give some more details of the problems they cause. Check back to make sure everything is spelt correctly.

Student answer 2 | Extract typical of a level ② answer

Watch your spelling, punctuation and grammar – answers will be marked on these.

> A poisnous gas thats called carbon monoxide is formed. Water is formed too. Carbon monoxide is called the silent killer. The solid that forms is soot. This gets made when there is not enough air getting to the fuel. Carbon dioxide formed in complete combustion causes the greenhouse effect.

This is quite a good description of why incomplete combustion takes place.

This mentions an effect of carbon dioxide produced by complete combustion.

This is good because it identifies both products.

Examiner summary
This answer includes both products and some of the problems they have. To improve on this, you could mention oxygen rather than air. It could also include more details about why incomplete combustion takes place.

A good answer identifies both substances.

During combustion of methane, carbon dioxide forms. This is the main greenhouse gas that causes climate change. In incomplete combustion there is not enough oxygen for methane to burn properly. It forms carbon monoxide and soot. Soot is carbon, and it makes polluting smoke. Carbon monoxide is a poisonous gas that can kill people.

This gives the reason why incomplete combustion occurs, but it could mention oxygen to improve the answer.

Here the answer identifies some of the problems caused by the substances.

Examiner summary

This is a good answer because it gives all the information asked for in the question. It names both the products, explains why incomplete combustion takes place and mentions some of the problems associated with the substances. It is clear and easy to read, with good spelling, grammar and punctuation.

  ResultsPlus

Move from level ① to level ②

This asks for two answers – information on complete and incomplete combustion. To progress to Level 2, you need to give some information about both types of combustion.

Move from level ② to level ③

To get an excellent mark, make sure your answer is clear and well organised. Answer all parts of the question in the order they are asked. Use scientific words, like climate change and oxygen correctly.

Here are three student answers to the following question. Read the answers together with the examiner comments around and after them.

Question — Biofuels

Grade D–A*

Biofuels are increasingly used as fuels for cars. Some biofuels are made from plants. The biofuel ethanol is made from sugar cane or sugar beet.

Petrol is a fossil fuel that is made from crude oil.

Discuss the advantages and disadvantages of using biofuels instead of petrol. (6)

Student answer 1 — Extract typical of a level ① answer

Explaining why biofuels are better for the environment would earn more marks.

This is a good disadvantage.

> Biofules are good for the environment. Burning petrol gives off carbon dioxide but burning biofuels doesn't. but if you have a biofule car it is hard to find a fuel station

The point about petrol is correct. However, when biofuels are burnt they do release carbon dioxide, but it is the same amount of carbon dioxide that was absorbed when the plants that made the biofuels grew.

Examiner summary

The answer gives a disadvantage of petrol and an advantage and disadvantage of a biofuel. The carbon-neutral nature of biofuels hasn't been explained clearly (or it was thought incorrectly that carbon dioxide is not a product of the combustion of biofuels). Not many scientific terms were used. Be careful with spelling scientific words, e.g., biofuel, and the last sentence should start with a capital letter and end with a full stop.

Student answer 2 — Extract typical of a level ② answer

This is the correct idea that no net carbon dioxide is evolved. It could be improved by mentioning photosynthesis in the answer.

> When you burn a biofuel it gives off carbon dioxide but this was originally absorbed when the plant grew. It is carbon neutralised.
> And we can grow more plants to make biofuels but petrol is running out.
> There are not many biofuel cars and it is difficult to find a place to fill them up.

More marks are awarded for using scientific terms. The correct term here is carbon-neutral.

Examiner summary

The advantages of biofuels are given, and also a disadvantage of petrol. However, these have not been stated using correct scientific terms, and in particular, carbon neutralised, is incorrect. A disadvantage of biofuels has also been given. The answer could have included further detail and additional advantages or disadvantages.

The advantage and disadvantage of growing plants for biofuels have been neatly contrasted. The answer could also mention the fact that forests have been chopped down to clear land for crop growth.

Looking ahead to future developments is useful.

> Biofuels are made from plants. As the plants grow they absorb carbon dioxide through photosynthesis. When the biofuel is burnt, carbon dioxide is released. Overall, the biofuels are carbon-neutral. However, burning petrol releases carbon dioxide, which leads to the greenhouse effect and global warming.
>
> As many plants as are needed can be grown to supply biofuels, but petrol is finite and supplies are running out. However, land is required to grow these plants, and this land can no longer be used to grow crops to feed people.
>
> Finally, there are many models of petrol cars available, but only a few of biofuel cars, and whilst there are many petrol filling stations, biofuel is not easily available. More biofuel filling stations will be available in the future.

The idea of carbon-neutral fuel has been very clearly explained here using scientific terms.

The consequences of carbon dioxide release have been mentioned here. The answer could be improved by explaining why the greenhouse effect might cause serious problems.

Examiner summary

The answer is a detailed discussion of the points. They have not just been listed, but advantages and disadvantages have been contrasted. Good scientific terms have been used, and the answer shows a clear understanding of ideas such as carbon neutrality. The answer is well organised into paragraphs, and the language is clear.

ResultsPlus

Move from level ① to level ②

The answer should give as many advantages and disadvantages of biofuels as possible. As the question says 'instead of petrol', these should be on the basis of replacing petrol with a biofuel. Use scientific terms when explaining the answer, e.g. non-renewable, carbon-neutral, combustion. A level 1 answer might explain some advantages or disadvantages so to move to level 2, more advantages and disadvantages are needed.

The answer should be well organised. One approach would be giving all of the advantages and then all of the disadvantages of a biofuel. Take care with spelling, especially for scientific terms, and write the answer in full sentences.

Move from level ② to level ③

A level 2 answer might have a list of advantages and disadvantages of both biofuels and petrol. To move to level 3, try to contrast the two fuels in the answer. For example, 'burning both petrol and biofuels releases carbon dioxide, but in the case of biofuels this carbon dioxide was recently absorbed when the plants that the biofuel is made from grew.' Make sure that you use as much scientific terminology as possible, for example, when explaining why biofuels are carbon-neutral refer to photosynthesis in plants as well as products of combustion.

Ensure that you 'discuss' rather than just list your points. Organise your answer into a logical sequence and make sure that the rules of spelling, punctuation and grammar are followed.

Here are three student answers to the following question. Read the answers together with the examiner comments around and after them.

Question — Polymers

Grade | D–A*

Alkenes can be used to make polymers (plastics).
Polymers are useful materials.

Used polymers can be recycled, burned or buried in landfill sites. Evaluate these methods of dealing with used polymers. (6)

Student answer 1 — Extract typical of a level ① answer

There is a spelling mistake in gasses, and it would be better to use a more scientific term such as toxic gases.

Be sure to spell ordinary words correctly, as well as scientific terms.

> It is good to recycle plastics because they can be used again. Burning plastic makes a nasty smell and gives dangerous gasses. Plastics do not rott in landfill sytes.

A good advantage has been given for recycling, and the answer goes on to mention all three methods. It's always useful to check back to the question and make sure everything has been answered.

Examiner summary

This answer has stated only one advantage or one disadvantage of each method. Scientific terms haven't been used – try to include these. There are some spelling errors.

Student answer 2 — Extract typical of a level ② answer

The answer bad for the environment does not explain why. The answer could be improved by giving reasons, e.g. natural habitats are destroyed.

> Recycling plastics does not waste the plastic. You have to put your plastics in recycling and some people cant be bothered. Burning a plastic can give off toxic fumes and cause pollution. But burning gives very little waste left at the end. Plastics are non-biodegradable and fill up landfill sites. Landfill sites are a mess and look very ugly. They are bad for the environment.

For recycling and burning, both an advantage and a disadvantage have been mentioned. It's always worth considering plus and minus points.

Examiner summary

The answer gives advantages and disadvantages of two of the disposal methods, but not for landfill disposal. It is important to go through each of the points in a question to be sure nothing is left out. There are more scientific terms than in sample answer 1 – toxic and non-biodegradable – which is good. The answer is organised and there is only one punctuation error (cant should be can't).

The advantage of recycling is carefully explained, linking it back to the raw material (crude oil) used to make the plastic.

The answer explains that not all points are clear cut – recently developed plastic bags can biodegrade. It is good to show that science is progressing.

Plastics are made from crude oil, a non-renewable resource that is dwindling, so it is important to recycle plastics and conserve this resource. However, the plastics must be sorted into types and the public has to cooperate in putting waste into recycling bins. Although burning plastics leaves little residue and can produce energy, toxic fumes can be produced and incinerators are unpopular with local residents. Plastics are non-biodegradable (although some types do decompose quickly) and fill up valuable landfill sites, although this method does not require sorting of the plastic and avoids the emission of toxic fumes. Overall, the best method is recycling as we must conserve our energy supplies.

A neat contrast has been made between an advantage and disadvantage of burning. The English is good, which will earn good marks.

The question asks to evaluate, so it is important to include a conclusion.

Examiner summary
The answer clearly explains the advantages and disadvantages of each method. The methods are compared with each other and a conclusion is drawn. A good range of scientific terms are used, and the answer is expressed well, with good spelling, punctuation and grammar.

 ResultsPlus

Move from level ❶ to level ❷

A level 1 answer might include advantages or disadvantages of recycling, burning and burying plastics in landfill sites. To move to level 2, make sure that all of the methods are covered. Each point should be explained as carefully as possible. Check that you have used scientific terms in your answer. Take care when spelling difficult words, and make sure you answer in complete sentences with capital letters and full stops.

Move from level ❷ to level ❸

A level 2 answer will probably mention all of the methods. To move to level 3, consider both advantages and disadvantages of each method. When looking at each method, try to contrast the advantages and disadvantages. Each one should be backed up by detailed scientific knowledge, and mention when more modern advances in science have altered the argument, for example with biodegradable plastics.

Your language must be written in a scientific way, justifying your statements with evidence and your answer should be fluent. In an 'evaluate', question, a conclusion is important, in this case by picking out the most important factor used to help choose the most favoured disposal method. There is not one correct answer here – but you have to explain why you picked the one you did.

Be the Examiner

Results Plus
Exam question report

Iron can be obtained from iron(III) oxide. Three mixtures were heated. (1)

1 carbon and iron(III) oxide
2 copper and iron(III) oxide
3 magnesium and iron(III) oxide

Iron would be formed in mixture:

A 1 only   **B** 2 only   **C** 1 and 3 only   **D** 1, 2 and 3

Answer: The correct answer is C.

How students answered

Iron would be formed in mixture 1, because carbon is more reactive than iron (this is how iron is extracted from its ore). Most students understood this, and put A as their answer. Less than half of students realised that mixture 3 will also react, because magnesium is more reactive than iron.

Results Plus
Exam question report

A water heater uses liquefied petroleum gas, LPG, which contains propane, C_3H_8.
Which equation correctly represents the complete combustion of propane? (1)

A $C_3H_8 + 4O_2 \rightarrow 3C + 4H_2O$ **B** $2C_3H_8 + 7O_2 \rightarrow 6CO + 8H_2O$ **C** $C_3H_8 + 3O_2 \rightarrow 3CO_2 + 4H_2$ **D** $C_3H_8 + 5O_2 \rightarrow 3CO_2 + 4H_2O$

Answer: The correct answer is D.

How students answered

Some students got this wrong, even though you don't need to know anything about balancing equations to get it right. If you look at the right-hand side of each equation, you'll see that D is the only one that has both carbon dioxide and water as products of the reaction.

Results Plus
Exam question report

Two fuels that can be used for cars are petrol and diesel. Which of the following statements is *false*? (1)

A The waste products produced during complete combustion are the same for both fuels.

B Loss of diesel from the fuel tank by evaporation in warm weather would be greater than for petrol.

C Diesel is more likely to solidify in very cold weather than petrol.

D Diesel is more difficult to ignite than petrol.

Answer: The correct answer is B.

How students answered

0 marks

Well over half of students got this wrong.

1 mark

Diesel fuel consists of longer carbon chains than petrol, so it has a higher boiling point and evaporates *less* easily. C and D are therefore correct and B is false. A is also correct.

Which of the following, present in the atmosphere, is least likely to contribute to global warming? (1)

A carbon dioxide   **B** hydrogen   **C** water vapour   **D** methane

Answer: The correct answer is B.

How students answered

0 marks

Nearly three-quarters of students got this one wrong.

1 mark

You should know that carbon dioxide, methane and water vapour are 'greenhouse gases' that are thought to contribute to global warming. Hydrogen is a reactive gas and will react with oxygen to form water.

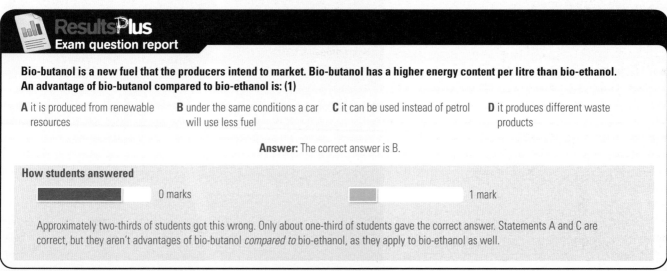

Bio-butanol is a new fuel that the producers intend to market. Bio-butanol has a higher energy content per litre than bio-ethanol. An advantage of bio-butanol compared to bio-ethanol is: (1)

A it is produced from renewable resources   **B** under the same conditions a car will use less fuel   **C** it can be used instead of petrol   **D** it produces different waste products

Answer: The correct answer is B.

How students answered

0 marks

1 mark

Approximately two-thirds of students got this wrong. Only about one-third of students gave the correct answer. Statements A and C are correct, but they aren't advantages of bio-butanol *compared to* bio-ethanol, as they apply to bio-ethanol as well.

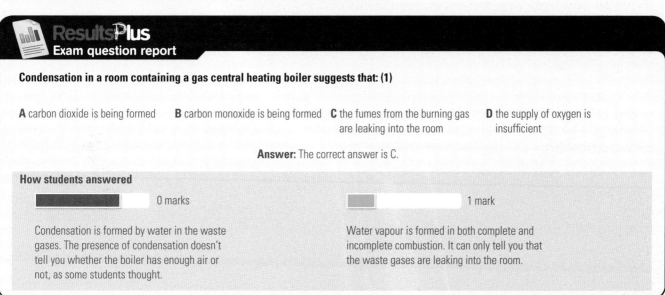

Condensation in a room containing a gas central heating boiler suggests that: (1)

A carbon dioxide is being formed   **B** carbon monoxide is being formed   **C** the fumes from the burning gas are leaking into the room   **D** the supply of oxygen is insufficient

Answer: The correct answer is C.

How students answered

0 marks

Condensation is formed by water in the waste gases. The presence of condensation doesn't tell you whether the boiler has enough air or not, as some students thought.

1 mark

Water vapour is formed in both complete and incomplete combustion. It can only tell you that the waste gases are leaking into the room.

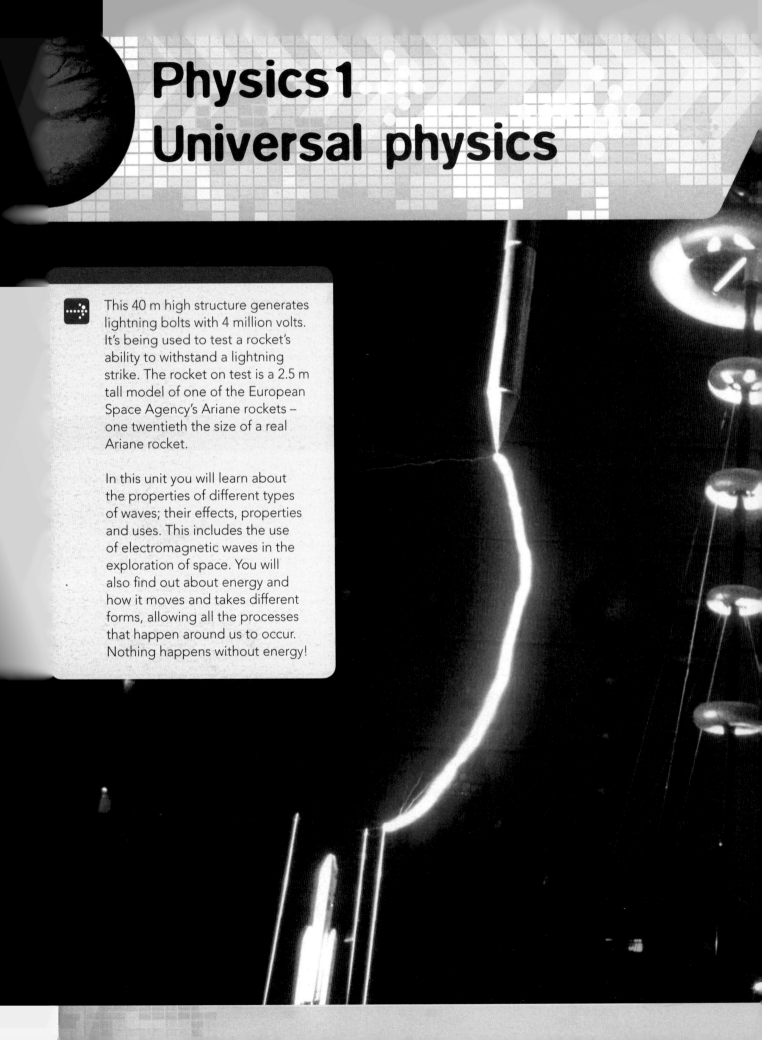

Physics 1
Universal physics

This 40 m high structure generates lightning bolts with 4 million volts. It's being used to test a rocket's ability to withstand a lightning strike. The rocket on test is a 2.5 m tall model of one of the European Space Agency's Ariane rockets – one twentieth the size of a real Ariane rocket.

In this unit you will learn about the properties of different types of waves; their effects, properties and uses. This includes the use of electromagnetic waves in the exploration of space. You will also find out about energy and how it moves and takes different forms, allowing all the processes that happen around us to occur. Nothing happens without energy!

Throughout this unit you will be required to:

0.1   Use equations given in this unit, or in a given alternate form

Ⓗ *0.2*   Use and rearrange equations given in this unit

0.3   Demonstrate an understanding of which units are required in equations

How do we know what is in the Universe?

ncient cultures from all over the world made up stories about the stars and planets. They also used our closest star (the Sun) to tell the time.

Shadows form a snake on the two days in a year when the length of night is equal to the length of daylight.

A *Twice a year, the Pyramid of Kukulcan creates shadows tha look like a giant snake. The snake was an important symbo the Mayan people who built the pyramid over 1000 years a*

1 What is a geocentric model? ❓

ResultsPlus
Watch Out!

Geocentric (Earth-centred) and *heliocentric* (Sun-centred) are ideas about the Solar System that have changed over time.

Like the ancient Mayans in Central America, the ancient Greeks made detailed measurements of the movements of objects in the sky. The Greek astronomer Ptolemy (c. 90–168) used these measurements to explain how the Sun, the Moon and the planets moved in **orbits**. His idea put the Earth in the centre of everything – a **geocentric** model.

The Polish astronomer Nicolaus Copernicus (1473–1543) thought that Ptolemy's measurements fitted a different model – a model with the Sun at the centre of the Solar System. The Church did not like this **heliocentric** model and a priest inserted an introduction into Copernicus's book before it was published saying that the book contained ideas with no proof.

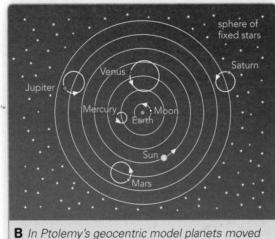

B *In Ptolemy's geocentric model planets moved in small circles as they orbited the Earth.*

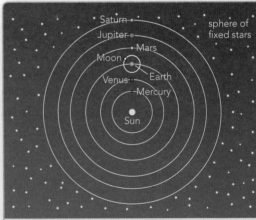

C *Copernicus's heliocentric model of the Solar System*

The **telescope** was invented at the end of the 16th century. It allowed scientists to see objects in space in much greater detail than with the **naked eye** and to find new objects.

Using a telescope, the Italian astronomer Galileo Galilei (1564–1642) discovered four of Jupiter's moons. By plotting their movements he showed that not everything orbited the Earth. This and other observations led him to support Copernicus's idea. However, the Church was still against the idea, and Galileo was put under house arrest for the last 10 years of his life.

2 a Describe two differences between Ptolemy's and Copernicus's models.
b Describe two similarities between the two models.
c Suggest one way in which our current model of the Solar System is different from Copernicus's.

3 How do Galileo's observations of Jupiter's moons support Copernicus's theory?

As telescopes improved, so more discoveries were made, including the planets Uranus and Neptune and the dwarf planet Pluto.

Astronomy today

Luminous objects in space give out **visible light** that travels as waves of energy. These visible **light waves** allow people to study distant objects. The invention of photography has allowed astronomers to make more detailed observations than they could by just making drawings.

Many objects in space do not give out much visible light but give out other types of energy-carrying waves, like **radio waves** and **microwaves**. Today, different types of telescopes are used to detect these different types of waves and scientists analyse the data collected to make conclusions about the Universe.

Skills spotlight

When someone proposes a new theory, other scientists need to check it to see if it is correct. How do you think Galileo found out about Copernicus's theory? How did Galileo check Copernicus's theory?

D *The Planck space telescope detects microwaves.*

4 Name three types of waves that telescopes can detect.

5 Suggest one reason why modern telescopes are put into space.

6 Describe how our understanding of the Universe has changed over time because of changes in technology.

Learning Outcomes

1.1 Describe how ideas about the structure of the Solar System have changed over time, including the change from the geocentric to the heliocentric models and the discovery of new planets

1.2 Demonstrate an understanding of how scientists use waves to find out information about our Universe, including:
a the Solar System b the Milky Way

1.3 Discuss how Galileo's observations of Jupiter, using the telescope, provided evidence for the heliocentric model of the Solar System

1.4 Compare methods of observing the Universe using visible light, including the naked-eye, photography and telescopes

HSW 14 Describe how scientists share data and discuss new ideas, and how over time this process helps to reduce uncertainties and revise scientific theories

>>>>>>>>>>>>>>>>>>>>>>>>>>>>

Why does a telescope have two lenses?

How can telescopes make faraway objects appear larger?

One of the first people to observe the night sky with a telescope was an English mathematician called Thomas Harriot (1560–1621). The telescope allowed him to see the Moon in much greater detail, and he made the first drawings of its craters.

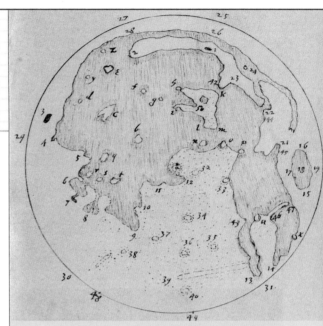

A *Thomas Harriot's earliest Moon drawing from 1609.*

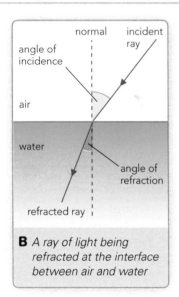

B *A ray of light being refracted at the interface between air and water*

Refraction

Light travels in straight lines. However, it can change direction when it moves into a different material. The change in direction is called **refraction** and happens at the **interface** (boundary) between the two materials. Figure B shows an example of this when light moves from air into water. A line at right angles to the interface is called the **normal** line.

Refraction in lenses

A **lens** is a transparent block that has been shaped so that its interface changes the directions of parallel light waves.

A **converging** (or **convex**) **lens** is a glass block that is curved on both sides to make it thicker in the middle. Light rays entering a lens from air are brought together or **converge**. Rays of light from distant objects are almost parallel when they reach us. A convex lens focuses these rays. The distance between the focus and the lens is called the **focal length** of that lens. The focal length of a converging lens can be found by focusing the **image** of a distant object onto a piece of paper.

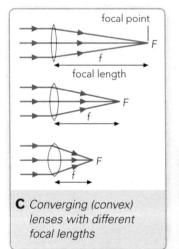

C *Converging (convex) lenses with different focal lengths*

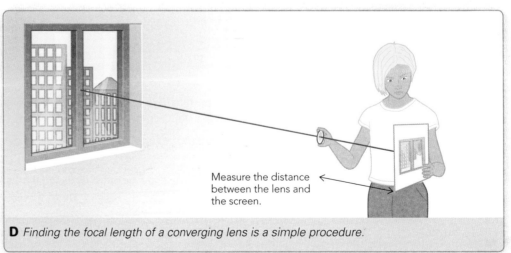

Measure the distance between the lens and the screen.

D *Finding the focal length of a converging lens is a simple procedure.*

1 Give examples of two different materials.

2 Write a short definition of 'refraction'.

In a **refracting telescope**, a convex lens (the **objective lens**) creates an image inside the tube and another lens (the **eyepiece lens**) is used to **magnify** this image.

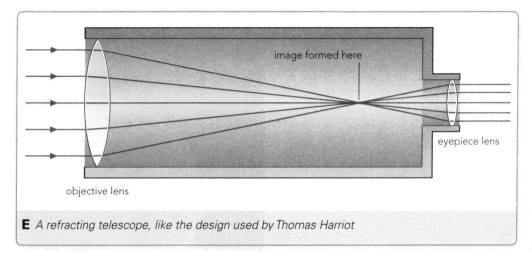

image formed here

eyepiece lens

objective lens

E *A refracting telescope, like the design used by Thomas Harriot*

 Refraction occurs because light travels at different speeds in different materials. Figure F shows how this causes a change in direction. Moving from air to glass, light slows down and refracts towards the normal. For light waves travelling in the opposite direction, the opposite change in speed occurs.

The waves are travelling in this direction.

This part of the wave slows down first.

air

glass or water

The waves end up travelling in this direction.

The waves travel more slowly in this medium.

F *Waves change speed at an interface, so they change direction.*

3 What does 'focal length' mean?

4 How can you measure the focal length of a convex lens?

Skills spotlight

Scientists use conventions (standard ways of doing things) in diagrams to help them to explain how things work. Describe two conventions that are used in the diagrams on these two pages.

Results Plus
Watch Out!

When you draw light rays, they must be straight (use a ruler!) and should have arrows on the lines to show the direction the light is travelling in.

 5 Describe what would happen if a light ray travelling inside a glass block entered a material where the speed was:
a greater than inside the glass **b** slower than inside the glass.

6 Explain how lenses work and how they are used in refracting telescopes.

Learning Outcomes

1.5 Explain how to measure the focal length of a converging lens using a distant object

1.8 Explain how the eyepiece of a simple telescope magnifies the image of a distant object produced by the objective lens (ray diagrams are not necessary)

1.10 Recall that waves are refracted at boundaries between different materials

(H) 1.11 Explain how waves will be refracted at a boundary in terms of the change of speed and direction

HSW 11 Present information using scientific conventions and symbols

P1.3　Lenses

What affects the image created by a converging lens?

If you hold a ruler at arm's length and look at the full Moon in the night sky, it will appear to be about 0.8 cm across. If you used Galileo's telescope, it would look twenty times bigger. In order to build his telescope so that it would magnify by twenty times, Galileo had to know what affected the image produced by his converging lenses.

Many optical devices use converging lenses to produce images that are **magnified**. The magnification of converging lenses depends on how curved their surfaces are and how close together they are placed. Some images can be shown on a screen. In order to make sketches of his observations, it is likely that Galileo would have allowed the image to fall on a screen so he could draw it. This is because he had produced a **real image** (an image in which the rays of light actually meet at the point where the image is seen).

This is different from a **virtual image**, like that seen in a mirror. In a virtual image, the rays of light appear to come from an image but do not actually come from that image. A virtual image cannot be shown on a screen.

A Galileo's 20x magnification telescope helped him dev the heliocentric model of the Solar System.

Your task

You are going to plan an investigation to find out how the distance between the object and the lens affects the image size. The hypothesis is that the closer an object is to a converging lens, the greater its magnified image will be. Your teacher will provide you with some materials to help you organise this task.

Learning Outcomes

1.6 Investigate the behaviour of converging lenses, including real and virtual images

1.7 Investigate the use converging lenses to: **a** measure the focal length using a distant object **b** investigate factors which affect the magnification of a converging lens (formulae are not needed)

When completing an investigation like this, one of the skills you will be assessed on is your ability to *assess the quality of the evidence*. There are 4 marks available for this skill. Here are two student extracts that focus on this skill. Other skills that you need for the assessment are dealt with in other lessons.

Student extract 1 — A basic response for this skill

This is a clear comment about the quality of the primary evidence.

> I think that there are some anomalous results in the data I collected when I did my practical. I have highlighted these in my results table. I think that this was because I did not measure the height of the image properly. It was quite hard to see sometimes. The secondary evidence I collected did not have any anomalies.

Difficulties with the method are important but should be discussed in a later section.

It is a good idea to state if there are anomalies or not.

Student extract 2 — A good response for this skill

A better way to do this would be to plot the anomalies and then do a line of best fit on your graph. This gives the full picture of your results.

> There are anomalies in the results collected from my practical. I have highlighted these on my results table. There were no anomalies in my secondary evidence. I have not included the anonamalous results when I did my graph or my calculations. They would not fit on my straight line .

Don't get rid of anomalous results completely. Put them in your results table but highlight which ones they are.

You need to explain why it is important to exclude anomalous results.

Check your spelling!

Results Plus

To access 2 marks
- Comment on the quality of your primary and secondary evidence
- Deal with any anomalies appropriately
- Say if you do not think there are any anomalies in your evidence

To access 4 marks
- Take account of any anomalies in your primary and secondary evidence
- Explain any adjustments you need to make to your evidence
- If you do not think there are anomalies, explain this and say that you are using all your evidence

How big can we make telescopes?

The largest refracting telescope ever made was built for an exhibition in Paris in 1900. It was 60 m long and had an objective lens that was 1.25 m in diameter. It was dismantled after the exhibition but the lenses can still be seen on display in Paris.

ResultsPlus
Watch Out!

Light can be reflected at the boundaries between different materials and/or pass through and be refracted.

Reflection in telescopes

Waves are also **reflected** at boundaries between different materials. This means that whenever light passes through a lens, some is reflected. This makes the image fainter. If a star is already very faint, this can be a problem for refracting telescopes.

Refracting telescopes also need to be very long to have large **magnifications**. Large lenses improve the magnification but are very heavy and are difficult to make in a perfect shape, meaning that images have distorted colours. These are not problems in a **reflecting telescope**, which has a curved mirror instead of an objective lens.

A *The giant refracting telescope of the 1900 Paris exhibition*

1 State two things that occur when light waves reach a boundary between different materials.

2 Why would a telescope using an objective lens be likely to produce a fainter image of a star than one that uses a mirror?

B *When light waves enter a glass block, some light is reflected.*

The curved **primary mirror** focuses parallel light rays from a distant object to an image in the same way as the objective lens in a refracting telescope. This image is then magnified by the eyepiece lens, also just like a refracting telescope. The first reflecting telescope that worked well was produced in 1668 by Sir Isaac Newton (1642–1727).

The majority of modern, large telescopes are reflecting. To view very faint, distant stars, a telescope needs to be able to collect tiny amounts of light. This means it needs to have a large diameter primary mirror to allow as much light to enter as possible.

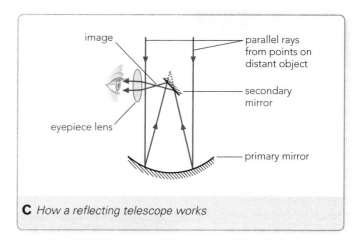

C *How a reflecting telescope works*

D *The primary mirror in the Gemini North telescope has a diameter of 8.1 m and a mass of 22 200 kg.*

Skills spotlight

Scientists need to be able to communicate the benefits of their work clearly to the public. You are the director of the Gemini telescope project. A journalist has asked you to justify the huge cost of building the telescope. Write a bullet point list explaining the scientific reasons why a larger diameter mirror is better and why it needs to be a reflecting telescope.

3 Why does having an inverted image not matter when a telescope is used in astronomy?

4 Why can reflecting telescopes be shorter than refracting ones that have the same magnification? (*Hint*: Compare Figure C on this page with Figure E on P1.2.)

5 Describe the similarities and differences between the way reflecting and refracting telescopes work.

Learning Outcomes

1.10 Recall that waves are reflected at boundaries between different materials

1.8 Explain how the eyepiece of a simple telescope magnifies the image of a distant object produced by the objective lens (ray diagrams are not necessary)

1.9 Describe how a reflecting telescope works

HSW **11** Present information, develop an argument and draw a conclusion, using scientific, technical and mathematical language, and ICT tools

What makes cliffs erode?

This house is about to collapse into the sea because the ground on which it stands has been worn away by the waves.

A *Sea waves carry energy which can wear away a cliff.*

> **1** If the water in a wave moved in the same direction as the energy, what would happen to the water in a swimming pool if you made waves at one end?

Sea waves **transfer** (carry) energy to the shore. When waves hit a cliff, the energy is transferred to the cliff and can wear it away. Water particles just move up and down as a wave passes – they aren't carried to the shore. Waves in which particles move at right angles to the direction that the wave is going are called **transverse waves**.

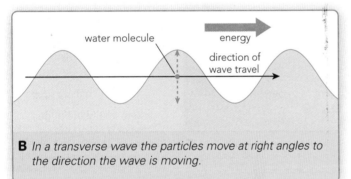

B *In a transverse wave the particles move at right angles to the direction the wave is moving.*

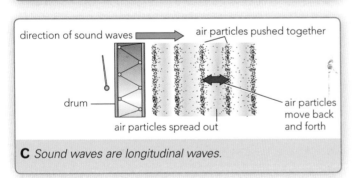

C *Sound waves are longitudinal waves.*

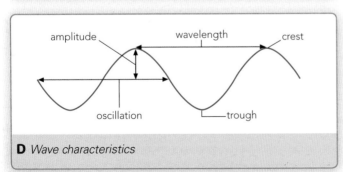

D *Wave characteristics*

Electromagnetic waves (such as light, radio waves, microwaves) are transverse and do not need a medium through which to travel.

Sound waves are not transverse. In a sound wave the particles move back and forth in line with the direction that the wave is going. These are **longitudinal waves**.

Earthquakes and explosions produce **seismic waves** that travel through the Earth. Solid rock material can be pushed and pulled (longitudinal seismic waves) or moved up and down or side to side (transverse seismic waves).

Frequency
Wave **frequency** is the number of waves passing a point each second. It is measured in **hertz (Hz)**. A frequency of 1 hertz means 1 wave passing per second.

Wavelength
The **wavelength** of a wave is the distance from a point on one wave to the same point on the next wave, measured in metres.

Amplitude
The **amplitude** of a wave is the maximum distance of a point on the wave from its rest position, measured in metres. Changes in the frequency, wavelength or amplitude of a wave can be used to carry information from one place to another.

Wave speed

How fast the energy in a wave travels is the **wave speed**. There are two ways to work this out.

$$\text{wave speed (metre/second, m/s)} = \frac{\text{distance (metre, m)}}{\text{time (second, s)}}$$

We can write this as symbols:

$$v = \frac{x}{t}$$

e.g.

For example, if a wave carries a surfer 52 metres in 8 seconds, the wave speed is:

$$\text{wave speed} = \frac{52}{8} = 6.5\,\text{m/s}$$

The wave speed is also linked to the wave frequency and wavelength:

wave speed (m/s) = frequency (Hz) × wavelength (m)

We can write this as symbols:

$$v = f \times \lambda$$

where λ (the Greek letter lambda) is the symbol for wavelength.

e.g.

For example, if some waves of 13-metre wavelength have a frequency of 0.5 Hz then the wave speed is:

$$v = 0.5 \times 13 = 6.5\,\text{m/s}$$

?

4 Calculate: **a** the speed of light waves, which travel 900 000 000 m in 3 seconds
b the speed of sound waves, which have a wavelength of 2 m and a frequency of 170 Hz
H **c** the wavelength of seismic waves which travel at 5000 m/s and have a frequency of 100 Hz.

H 5 Draw a diagram of a transverse wave to help you explain what is meant by amplitude and wavelength. Then explain what is meant by frequency and wave speed and how these are connected to the wavelength.

?

2 Give two examples of transverse waves and two examples of longitudinal waves.

3 Sea wave crests pass a stick twice every second. What is the frequency?

Skills spotlight

Scientists need to be able to write instructions for carrying out observations. Write a brief plan for a way to work out the wavelength of waves in the sea if you are at the end of a 50 m pier and have a stopwatch.

ResultsPlus
Watch Out!

The amplitude of a wave is from the middle to the top or bottom, *not* the distance from crest to trough.

Learning Outcomes

1.12 Describe that waves transfer energy and information without transferring matter

1.13 Use the terms of frequency, wavelength, amplitude and speed to describe waves

1.14 Differentiate between longitudinal and transverse waves by referring to sound, electromagnetic and seismic waves

1.15 Use of both the equations below for all waves:

wave speed (metre/second, m/s) = frequency (hertz, Hz) × wavelength (metre, m)

$$v = f \times \lambda$$

wave speed (metre/second, m/s) = distance (metre, m)/time (second, s)

$$v = \frac{x}{t}$$

HSW 5 Plan to test a scientific idea, answer a scientific question, or solve a scientific problem

Is there any light you can't see?

What are infrared and ultraviolet?

For many centuries ships at sea have signalled to each other using lights. The signals are flashes of light that form a code. Remote controls in your home also use this idea but they use flashing signals that you cannot see. They use flashes of infrared waves.

A *An infrared remote control seen on a digital camera screen. Many digital cameras can detect infrared waves but our eyes cannot.*

The discovery of infrared

William Herschel (1738–1822) was a British astronomer. He put dark, coloured filters on his telescope to help him observe the Sun safely. He noticed that different coloured filters heated up his telescope to different extents and he wondered whether the different colours of light contained different 'amounts of heat'.

To test his idea he used a prism to split sunlight into a spectrum and then put a thermometer in one of the colours. He placed two other thermometers either side of the spectrum. As he changed the colour from violet to red, he found that the temperature rose.

1 List these colours in order of how much they heat up a thermometer: blue, green, orange, red, violet, yellow.

2 Suggest why Herschel needed to use coloured filters on his telescope.

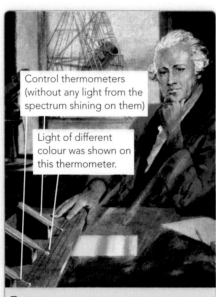

Control thermometers (without any light from the spectrum shining on them)

Light of different colour was shown on this thermometer.

B *William Herschel and his experiment*

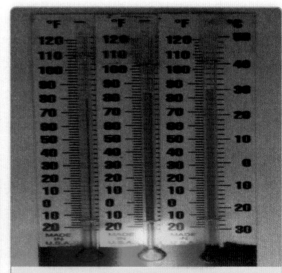

C *It is simple to set up Herschel's experiment today and see the same results.*

Herschel wondered what would happen if he measured the temperature just beyond the red end of the spectrum, where there was no visible light. He found that this gave the highest temperature. He had discovered **infrared** waves (infrared radiation, **IR**).

Going beyond violet

Johann Ritter (1776–1810) found out about Herschel's work and in 1801 set about trying to find 'invisible rays' at the other end of the spectrum. He used silver chloride, a chemical that breaks down to give a black colour when exposed to light.

It was already known that silver chloride turned black more quickly in violet light than in red light. Ritter showed that silver chloride turned black fastest when exposed to 'invisible rays' just beyond violet. These rays were later called **ultraviolet** waves (ultraviolet radiation, **UV**).

D A bird can see a flower using UV (left) but our eyes only detect visible light (right).

3 How might Herschel have continued his experiments?

4 How could Ritter have found out about Herschel's work?

Electromagnetic waves

Visible light, infrared and ultraviolet radiations are all types of **electromagnetic radiation**. The waves transfer energy from one place to another. The electromagnetic vibrations are at right angles to the direction in which the energy is being transferred by the wave (a bit like the water waves in Figure B on P1.5). So they are transverse waves.

5 Give two ways in which infrared radiation and ultraviolet radiation are similar types of wave.

6 Explain how scientists in the beginning of the 19th century knew that there is radiation that our eyes cannot detect beyond both ends of the visible spectrum.

Watch Out!

Infrared waves, infrared radiation and IR are all names for the same thing!

Learning Outcomes

2.1 Demonstrate an understanding of how Herschel and Ritter contributed to the discovery of waves outside the limits of the visible spectrum

2.2 Demonstrate an understanding that all electromagnetic waves are transverse

HSW **5** Plan to test a scientific idea, answer a scientific question, or solve a scientific problem by controlling relevant variables

What is the electromagnetic spectrum?

Sound waves need particles to vibrate in order to transfer energy. Sound cannot travel through space because there are no particles. Electromagnetic waves, such as light, can move through space, so we can see the Sun but we cannot hear any sound it makes.

1 The Moon is 400 000 km away. How long does it take light reflected from the Moon to reach your eyes?

ResultsPlus
Watch Out!

One way to help you remember the order of the seven groups of the electromagnetic spectrum is the three on the shorter side of visible light are more dangerous (UV, X-ray, gamma rays); the three on the longer side of light are used in the home (IR, microwaves, radio waves).

2 Name three different types of electromagnetic waves.

3 Which part of the electromagnetic spectrum has a higher frequency than X-rays?

4 What type of electromagnetic wave has a wavelength between visible light and X-rays?

The fastest speed there is

Electromagnetic waves can travel without any particles to vibrate. This means that they can move easily through a **vacuum**, such as space. All electromagnetic waves travel at 300 000 kilometres per second in a vacuum – this is the fastest speed anything can move.

The colour of visible light depends on its wavelength. If the wavelength of a light wave is longer than that of red light, human eyes cannot see it. Infrared (IR), microwave and radio waves have longer wavelengths than red light.

A *A poster from the film* Alien. *Notice the tagline.*

Electromagnetic waves with shorter wavelengths have higher frequencies. Ultraviolet radiation has a higher frequency than visible light. Even shorter wavelengths are present in **X-rays** and then **gamma rays**.

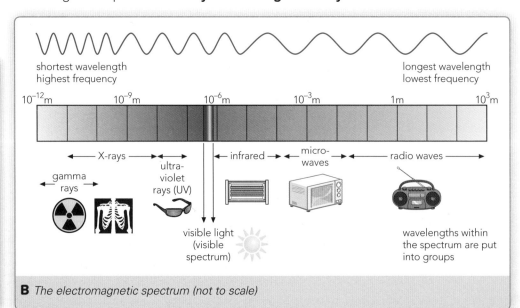

B *The electromagnetic spectrum (not to scale)*

The full range of electromagnetic waves is called the **electromagnetic spectrum**. It's like a bigger version of the visible spectrum and also includes all the wavelengths we cannot see. The spectrum is continuous so all values of wavelength are possible. It is convenient to group the spectrum into seven wavelength groups, as shown in Figure B.

Modern astronomy tries to observe stars and galaxies by detecting the various parts of the electromagnetic spectrum they give off. Radio astronomy and X-ray astronomy are examples that have developed relatively recently. The Hubble Space telescope can detect visible light, UV and IR.

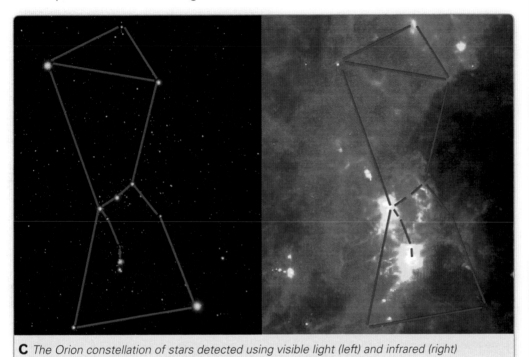

C *The Orion constellation of stars detected using visible light (left) and infrared (right)*

5 How do we know electromagnetic waves can travel through a vacuum, such as space?

6 List the seven parts of the electromagnetic spectrum in order and explain how the wavelength, frequency and speed in a vacuum change from one part to the next.

Learning Outcomes

2.2 Demonstrate an understanding that all electromagnetic waves are transverse and that they travel at the same speed in a vacuum

2.3 Describe the continuous electromagnetic spectrum including (in order) radio waves, microwaves, infrared, visible (including the colours of the visible spectrum), ultraviolet, X-rays and gamma rays

2.4 Demonstrate an understanding that the electromagnetic spectrum is continuous from radio waves to gamma rays, but the radiations within it can be grouped in order of decreasing wavelength and increasing frequency

HSW 9 Recall, analyse, interpret, apply and question scientific information or ideas

What are the dangers of electromagnetic waves?

In World War II, devices called magnetrons generated microwaves for radar systems to detect German war planes. In 1945, Percy Spencer noticed that a chocolate bar melted in his pocket when he stood in front of a working magnetron. Spencer used this discovery to construct the first microwave oven.

A The earliest microwave ovens cost $3000.

Dangers of electromagnetic waves

All waves transfer energy. A certain microwave frequency can heat water and this frequency is used in microwave ovens. This heating could be dangerous to people because our bodies are mostly water, so mobile phones, which also use microwaves, use different frequencies. Current scientific evidence tells us that, in normal use, mobile phone signals are not a health risk.

IR radiation is used in grills and toasters to cook food. Our skin absorbs IR, which we feel as heat. Too much infrared radiation can damage or destroy cells, causing burns to skin.

1 Why should you be careful not to stand too close to a bonfire?

2 Why do microwave ovens have shields in them to stop the waves escaping?

3 Describe two ways to protect yourself against skin damage by UV when in bright sunlight.

B Health agencies in many countries need to look for ways of warning people of the dangers of sun exposure. This photo is part of a publicity campaign from Australia to raise awareness of the dangers of melanoma, a type of skin cancer that can be caused by sun exposure. Australia has the highest incident of skin cancer in the world.

Higher-frequency waves transfer more energy and so are potentially more dangerous. Sunlight contains UV, which carries more energy than visible radiation. The energy transferred by UV to our cells can damage their **DNA**. Too much exposure to UV can damage skin cells so much that it leads to **skin cancer**.

The UV in sunlight can also damage our eyes. Skiers and mountaineers can suffer temporary 'snow blindness' because so much UV is reflected from snow.

D *This children's warning device changes colour depending upon how much UV it absorbs.*

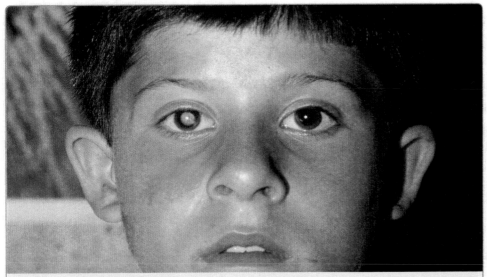

C *UV exposure to eyes over a long time can cause cataracts – a clouding of the lens of the eye and reducing vision.*

Skills spotlight

Look at the UV warning device in Figure D. Explain how using the Sun Sensor Watch can help prevent skin cancer.

Some parts of the body can absorb X-rays. X-rays are even higher energy than UV. Gamma rays carry even more energy than X-rays. X-rays and gamma rays can penetrate the body. Excessive exposure to X-rays or gamma rays may cause **mutations**, or changes, in DNA that may kill cells or cause cancer.

Results Plus
Watch Out!

Higher frequency waves have a shorter wavelength. Higher frequency means more energy, which means more danger.

4 Why do people have hospital X-ray photographs taken if X-rays are so dangerous?

5 Why is it useful to have UV predictions in the weather forecast?

6 Draw a table with a row for each part of the electromagnetic spectrum, with increasing frequency down the table. In the second column list any hazards to life that you know for each part of the spectrum. Use your table to explain how the frequency relates to the potential danger.

Learning Outcomes

2.5 Demonstrate an understanding that the potential danger associated with an electromagnetic wave increases with increasing frequency

2.6 Relate the harmful effects, to life, of excessive exposure to the frequency of the electromagnetic radiation, including:
   a microwaves: internal heating of body cells
   b infrared: skin burns
   c ultraviolet: damage to surface cells and eyes, leading to skin cancer and eye conditions
   d X-rays and gamma rays: mutation or damage to cells in the body

HSW 12 Describe the benefits, drawbacks and risks of using new scientific and technological developments

>>>>>>>>>>>>>>>>>>>>>>>>>

How can the electromagnetic spectrum fight crime?

Many banknotes have a pattern of yellow, green or orange circles. This 'EURion pattern' can be detected by software inside colour photocopiers and scanners which prevents them from copying the banknotes.

A *The coloured pattern of small circles on some banknotes prevents digital copying.*

Vision and photography

The EURion pattern on banknotes can be seen when **illuminated** because it reflects certain wavelengths of visible light. These wavelengths can be detected by our eyes and by photographic film, digital cameras, photocopiers and scanners.

Security

Other ways of stopping banknotes being counterfeited cannot always be seen. Some materials absorb UV radiation and re-emit it as visible light. This is called **fluorescence.** It is used for security markings on property and banknotes, which can be checked with a UV lamp. Some security lights use **fluorescent lamps**. These also produce UV waves and use a fluorescent material on the inside of the bulb glass.

X-ray scanners are used in airports to detect objects hidden on the body as well as in luggage.

All warm objects give off some heat as IR radiation. CCTV cameras that detect IR are used to watch people at night. IR radiation can pass through fog, making **thermal imaging** useful in daytime too.

B *Forged banknotes can be detected using UV light because they do not have markings that glow. This is a real note.*

1 Suggest why fluorescent lamps have fluorescent materials on the inside surface of the bulb glass.

2 Why can't suspects avoid police thermal imaging cameras by hiding in bushes?

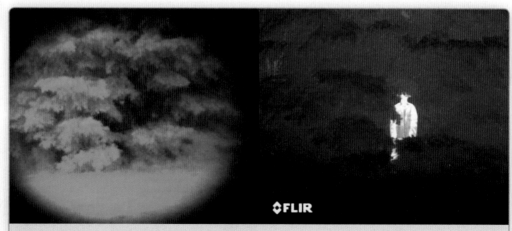

C *Both of these photos show the same scene. On the right a suspect is caught by thermal imaging.*

Communications

Radio waves are widely used for broadcasting and communications. Microwaves and radio waves carry TV signals, including those from satellites. Wi-Fi™ wireless network connections for computers also use radio waves. Microwaves carry mobile phone signals.

IR waves carry signals a short distance from remote controls to devices like TVs. IR signals are also sent down **optical fibre** cables for telephone and Internet communications.

Food and medicine

Gamma rays transfer a lot of energy which can kill cells. For this reason, they are used to sterilise food and surgical instruments by killing microorganisms. UV light also kills bacteria, so it can be used to disinfect water and sewage.

Gamma rays are used to kill cancer cells in **radiotherapy**. They can also be used to detect cancer. A chemical that emits gamma rays is injected into the blood. These can be detected by a scanner outside the body. The chemical is designed to collect inside cancer cells and a scanner then locates the cancer by finding the source of the gamma rays.

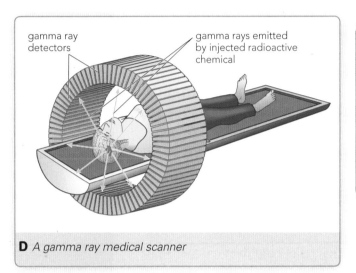

gamma ray detectors

gamma rays emitted by injected radioactive chemical

D A gamma ray medical scanner

ResultsPlus
Watch Out!

If you are asked to explain a use of an electromagnetic wave then try to include information about its frequency and wavelength.

3 Why would there be UV lamps at a sewage works?

4 Some people worry when they hear that their herbs have been treated with gamma rays. Why shouldn't they be concerned?

5 Describe two ways gamma rays can be used for medical purposes.

6 Draw a table with a row for each part of the electromagnetic spectrum, with increasing wavelength down the table. In the second column list any uses of each part of the spectrum, with a brief explanation of each application.

Learning Outcomes

2.7 Describe some uses of electromagnetic radiation
  a radio waves: including broadcasting, communications and satellite transmissions
  b microwaves: including cooking, communications and satellite transmissions
  c infrared: including cooking, thermal imaging, short range communications, optical fibres, television remote controls and security systems
  d visible light: including vision, photography and illumination
  e ultraviolet: including security marking, fluorescent lamps, detecting forged bank notes and disinfecting water
  f X-rays: including observing the internal structure of objects, airport security scanners and medical X-rays
  g gamma rays: including sterilising food and medical equipment, and the detection of cancer and its treatment

 13 Explain how and why decisions that raise ethical issues about the uses of science and technology are made

>>>>>>> Kryptonite is a fictional green-glowing element from *Superman* stories. Can any real elements glow?

What is ionising radiation?

In May 1962, the first edition of the science fiction comic book *Incredible Hulk* introduced physicist Dr Bruce Banner. In the book Dr Banner was working for the US Department of Defense to produce a 'gamma bomb'. He was accidentally exposed to a test explosion and became the Hulk. Fifty years later, the US Department of Defense is currently developing just such a bomb. But how would the actual effects of a gamma bomb explosion differ from the fictional ones described in the cartoon?

A *The Incredible Hulk was created by exposure to gamma rays*

Radiation and radioactivity

Gamma rays cause an increased risk of cancer by causing mutations in a cell's DNA. This is because gamma rays are **ionising radiation**. Such radiation can remove electrons from atoms to form **ions**. Ions are chemically very reactive. If atoms in the cell are ionised, the reactions that follow can damage the DNA. More exposure to gamma rays can cause the cell to be destroyed. Gamma rays can be used for sterilisation because they destroy bacterial cells.

1 Why are ionising radiations dangerous? (?)

Skills spotlight

A gamma bomb would emit huge amounts of energy in the form of gamma rays. It could kill people directly but would also cause long-term increased risks of cancer for the survivors. Write a paragraph to explain why one country's development of such a bomb might cause other countries to develop similar weapons, and whether you think that this is a good thing or not.

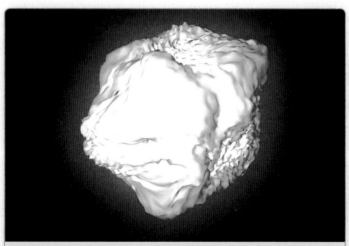

B *Radium metal glows naturally. Radium is radioactive and gives off energy, including gamma rays and visible light, all the time.*

Some elements such as radium naturally emit (give out) gamma (γ) waves all the time. Such elements are said to be **radioactive**. Not all radioactive substances emit gamma waves. Others emit particles, called **alpha (α) particles** and **beta (β) particles**. All three are types of ionising radiation and transfer energy from the radioactive material to their surroundings.

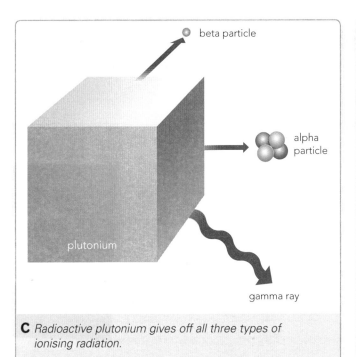

C *Radioactive plutonium gives off all three types of ionising radiation.*

D *Ionising radiation hazard symbol*

Alpha particles and beta particles are not electromagnetic radiation. They are particles of matter that are emitted with a lot of kinetic energy. This energy can ionise atoms. So, like gamma rays, alpha and beta particles can be hazardous to life as they can also damage cells and DNA within the cells.

2 What do the symbols α, β and γ mean?

3 How are alpha, beta and gamma radiations similar?

4 Give one difference between alpha particles and gamma rays.

5 A Geiger–Müller tube is a machine that can detect ionising radiations. Describe how a scientist could test whether a sample of radium emits radiation all the time.

6 Explain why gamma, UV and X-rays can all cause cancer but other parts of the electromagnetic spectrum do not. How do alpha and beta particles fit in with your explanation?

Results Plus
Watch Out!

UV and X-rays can cause cancer because they are also ionising radiations, so they can damage cells and DNA. However, they are not produced by radioactive materials.

Learning Outcomes

2.8 Recall that ionising radiations are emitted all the time by radioactive sources

2.9 Describe that ionising radiation includes alpha and beta particles and gamma rays and that they transfer energy

HSW **13** Describe the social, economic and environmental effects of decisions about the uses of science and technology

What's the longest way of writing your address you can think of?

The Universe

 How have ideas about our place in the Universe changed?

You might have written your address like the one in Figure A at some time. Observing electromagnetic radiation has allowed us to find out more about our exact location in the Universe.

N.E.Body
21 New Road
Oldtown
AB98 7CD
United Kingdom
Europe
Earth
Solar System
Milky Way
Universe

A *Your complete address!*

ResultsPlus
Watch Out!

The naked eye means making observations without using telescopes, binoculars, cameras or any other devices.

Ancient astronomers made detailed observations of the positions of the **stars** using the **naked eye**. They worked out that stars were further away than planets and thought that all the stars were fixed on a shell around the Earth, all at the same distance. They could also see patches of fuzzy light that they called **nebulae** (from the Latin word for clouds) and a band of light across the sky, which they named the **Milky Way**.

B *An image of the Orion nebula taken by the Hubble Space Telescope. The Orion nebula is a cloud of gas. The bright points in the image are stars.*

Using a telescope, Galileo Galilei (1564–1642) discovered that the Milky Way and some nebulae were actually groups of stars. He suggested that stars were other suns and were at different distances from the Earth.

As telescopes with greater magnifications were invented, astronomers made more discoveries about the **Solar System** and the stars. Astronomers in Galileo's time could only observe about six planets, but now we know that the Solar System contains eight planets in addition to some dwarf planets, many moons and lots of smaller bodies such as asteroids.

We also now know that the Sun is just one of millions of stars that make up our **galaxy** – the Milky Way. It is shaped like the galaxy shown in Figure A. There are billions of other galaxies. Some of the nebulae seen from Earth are actually other galaxies. All the galaxies make up the **Universe**.

Astronomers have also used telescopes to make observations that allow them to work out the sizes of planets, stars and galaxies and the distances between them. Figure C shows some sizes and distances.

1 Write down two discoveries made using telescopes.

2 Which is the nearest star to the Earth?

3 What is the Solar System?

4 What is: **a** the Milky Way **b** the Universe?

Object	Diameter (km)	Distance from Earth (km)
Earth	13 000	
Moon	3 500	384 000
Sun	1 400 000	150 000 000
Neptune (furthest planet from the Sun)	50 000	4 500 000 000
Proxima Centauri (nearest star to the Sun)	100 000	4.0×10^{13}
Milky Way	1.0×10^{18}	
Andromeda (nearest large galaxy to the Milky Way)	1.4×10^{18}	2.5×10^{19}
Most distant galaxy observed	too far to measure	1.2×10^{23}

C *Some sizes and distances in space*

Skills spotlight

Uncertainties in scientific knowledge change over time.
a Which measurements in Figure C might Galileo have known?
b Why could he not have known about the others?
c Which measurements are likely to change in the future? Explain your answer.

It is hard to imagine the sizes and distances in the Universe. Figures D and E show ways of thinking about the relative sizes and distances of the Earth, Moon and Sun.

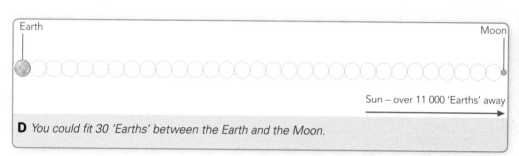

Earth　　　　　　　　　　　　　　　　　　　　　　Moon

Sun – over 11 000 'Earths' away

D *You could fit 30 'Earths' between the Earth and the Moon.*

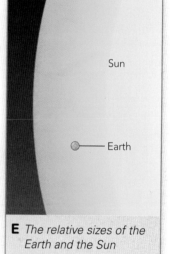

Sun

Earth

E *The relative sizes of the Earth and the Sun*

5 Write these in size order, starting with the smallest: Sun, Universe, Earth, Milky Way, Solar System, Moon.

6 Why would it be very difficult to draw a diagram similar to Figures D and E to help you to think about the relative sizes of a galaxy and the Earth?

7 Explain why the development of telescopes with better magnification has helped astronomers to find out more about the Universe.

Learning Outcomes

3.1 Recall that the Solar System is part of the Milky Way galaxy

3.2 Describe a galaxy as a collection of stars

3.3 Recall that the Universe includes all of the galaxies

3.4 Compare the relative sizes of and the distances between the Earth, the Moon, the planets, the Sun, galaxies and the Universe

HSW 14 Describe how scientists share data and discuss new ideas, and how over time his process helps to reduce uncertainties and revise scientific theories

P1.12 Spectrometers

···· **How do scientists study light from the Sun and stars?**

Helium is a light gas used to fill balloons. It was first discovered in the Sun in 1868. It was discovered by two scientists who were studying the light coming from the Sun. It was not found in Earth's atmosphere until nearly 30 years later.

B A CD can split light into the colours of the spectrum.

Light from the Sun is a mixture of different colours, forming a **spectrum**. However, gases in the Sun's atmosphere absorb some of the light in the spectrum and cause dark bands to form. The gases in the Earth's atmosphere can also cause dark bands to form in the Sun's spectrum.

Visible light is a mixture of light with different wavelengths and frequencies. These different wavelengths can be split up using a prism, or by something that has lots of fine lines on it, such as a CD or DVD. A device that contains something that can split up the different wavelengths of light is called a spectrometer. You can use a simple **spectrometer** to investigate light from the Sun.

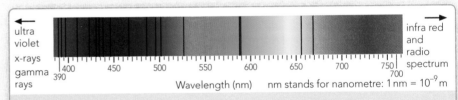

C The dark lines in this spectrum can help scientists to find out which elements are present in the atmosphere of the Sun.

Your task

You are going to plan an investigation that will allow you to find out about the way that sunlight is absorbed in different parts of the spectrum at different times of the day. The hypothesis is that the Sun's spectrum changes during the course of a day. Your teacher will provide you with some materials to help you organise this task.

Learning Outcomes

3.8 Construct a simple spectrometer, from a CD or DVD, and use it to analyse common light sources

When completing an investigation like this, one of the skills you will be assessed on is your ability to *draw a conclusion about your experiment.* There are 6 marks available for this skill. Here are two student extracts that focus on this skill. Other skills that you need for the assessment are dealt with in other lessons.

Student extract 1 | **A typical response for this skill**

You should include a comment on whether the hypothesis is supported by your conclusion or not.

> Both my pictures and the pictures I found on the Internet show that there are different lines in the spectrum at different times of the day. This is what our hypothesis is. The pictures show that there are more lines at sunset than at midday. This can be seen from the photos I took of the spectra.

This is good because it refers to the evidence.

Student extract 2 | **A good response for this skill**

This is a clear summary and uses evidence from both primary and secondary sources.

> Both my results and the photos I found on the Internet as secondary evidence show that there are more lines on the spectrum at sunset than there are at midday. This supports the hypothesis. I would expect this from my scientific knowledge which tells me that the sunlight will be absorbed as it passes through the atmosphere. The areas of the spectrum where the sunlight is absorbed will show up as dark bands. The Sun is at different angles in the sky at different times of day. In the morning and evening the light has had to pass through more of the atmosphere to reach us, so there is more chance of light being absorbed by gases in the atmosphere.

Good use of scientific knowledge to explain the conclusion.

 ResultsPlus

To access 4 marks

- Provide a conclusion based on all your collected evidence
- Explain your conclusion
- Describe any relevant mathematical relationships in your conclusion

To access 6 marks

You also need to:

- Refer to the original hypothesis in your conclusion
- Refer to relevant scientific ideas in your conclusion

How do astronomers use the electromagnetic spectrum to investigate the Universe?

The Sun appears to us as just a glowing ball in the sky. It looks much more interesting if you look at a photograph taken using a telescope that detects X-rays. Figure A is an X-ray image that shows huge explosions in the Sun's atmosphere called solar flares.

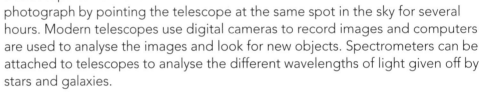

A

1 How does using a camera to record the image seen through a telescope allow astronomers to gather more data?

2 Look at Figure C. Write down two differences between radio telescopes and telescopes that detect visible light.

3 Write down two things that can be seen using X-rays that cannot be seen using visible light.

Early telescopes let people see objects that emitted **visible light**. When photography was invented the amount of data that could be gathered with these telescopes increased. Objects that are too faint to be seen by looking through a telescope can be recorded as a photograph by pointing the telescope at the same spot in the sky for several hours. Modern telescopes use digital cameras to record images and computers are used to analyse the images and look for new objects. Spectrometers can be attached to telescopes to analyse the different wavelengths of light given off by stars and galaxies.

Most objects that astronomers observe give out energy in all parts of the **electromagnetic spectrum**. Modern telescopes can be designed to detect almost any part of the electromagnetic spectrum, showing us things that can't be detected using visible light. The **X-ray** image in Figure B shows the remains of an exploded star (circled) that does not show up on the **ultraviolet** (UV) image or on images made using visible light.

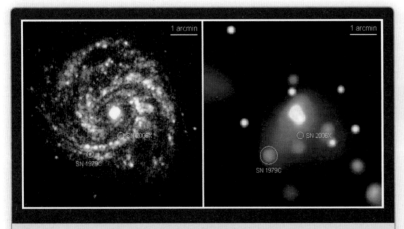

B *Images of the M100 galaxy in ultraviolet (left) and X-ray radiation (right). They were both taken by the Swift telescope.*

C *These telescopes detect radio waves.*

Telescopes in space

The Hubble Space Telescope has been in orbit around the Earth since 1990. It can produce very clear and sharp images of stars and galaxies because it is above the atmosphere. This means that light waves from space are not reflected or refracted by clouds and dust or even by movements of air in the atmosphere.

Other telescopes are put into orbit because the atmosphere absorbs the wavelength of the radiation they are designed to detect. If these telescopes were on Earth they would not detect anything.

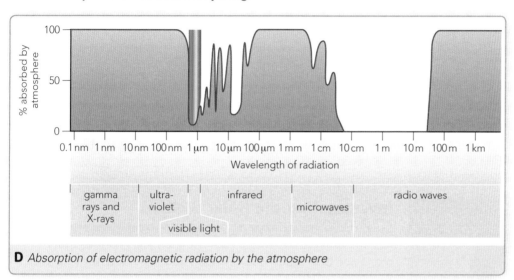

D *Absorption of electromagnetic radiation by the atmosphere*

4 Look at the photos in Figure B. Why does the Swift telescope need to be in orbit around the Earth?

 5 a Look at Figure D. Which of these must be put into orbit around the Earth to work?
i UV telescope
ii a telescope to detect microwaves
iii an optical (visible light) telescope
iv an IR telescope
b Use data from the graph to explain your answers to part a.

6 Describe three ways in which observations by present-day astronomers are different from the kinds of observations made by Galileo in the Middle Ages.

ResultsPlus
Watch Out!

More data has been collected and more objects have been discovered since the development of telescopes that detect more of the electromagnetic spectrum than visible light.

Skills spotlight

Science has been able to answer different questions at different times in history. What type of questions about the Universe:
a can astronomers answer now that Galileo could not?
b cannot be answered now but may be answered in the future?

Learning Outcomes

3.5 Describe the use of other regions of the electromagnetic spectrum by some modern telescopes

3.7 Demonstrate an understanding of the impact of data gathered by modern telescopes on our understanding of the Universe, including:
a the observation of galaxies because of improved magnification
b the discovery of objects not detectable using visible light
c the ability to collect more data

3.9 Explain why some telescopes are located outside the Earth's atmosphere

3.10 Analyse data provided to support the location of telescopes outside the Earth's atmosphere

HSW **4** Identify questions that science cannot currently answer, and explain why these questions cannot be answered

How do scientists look for evidence of life beyond the Earth?

Pioneer 10 and Pioneer 11 are space probes that were launched into space in the 1970s. They are now well beyond the Solar System, but they won't reach another star for about 2 million years. They both carry a message designed to show intelligent life where to find us (shown in Figure A).

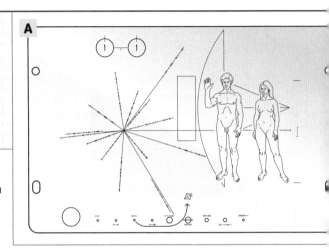

A

Earth is the only place where we know that life exists. Life may exist in many other places in the Universe, but at present we have no evidence to show that it does.

B *One of the Viking landers on Mars*

Investigating the Solar System

In 1976 two Viking **landers** touched down on Mars. They did experiments on the soil to look for chemical changes that may have been caused by living organisms. One of these soil experiments showed some changes but the others did not. Scientists decided that they had found no evidence for life.

All life as we know it depends on liquid water. Scientists are looking for evidence that there is or was water on planets and moons in the Solar System. **Space probes** orbiting Mars have photographed channels created by flowing water in the past. The *Phoenix* lander discovered frozen water in the Martian soil in 2008.

Rovers on Mars have also taken close-up photographs of the rocks and soil but have not found any evidence of life.

1 Why are scientists looking for evidence of water on Mars?

2 Is *Spirit* a space probe, a lander or a rover?

C Spirit *investigated the surface of Mars for over 6 years and covered over 12 miles.*

Beyond the Solar System

Scientists have discovered planets orbiting other stars. These are too far away for telescopes to produce clear images, but scientists can sometimes find out about the gases in the planet's atmosphere.

The Earth has a lot of oxygen in its atmosphere because plants release oxygen during photosynthesis. If oxygen is detected in the atmosphere of another planet, it would be evidence that life *may* exist on that planet.

Searching for intelligent life

Some of the electromagnetic radiation that we use for communications travels out into space and could be detected by intelligent life. Alien civilisations may even be sending out signals to try to contact other life forms. The **Search for Extraterrestrial Intelligence** (or **SETI**) is a name for a series of projects that analyse radio waves coming from space. They look for signals possibly produced by intelligent beings. No messages have been detected so far.

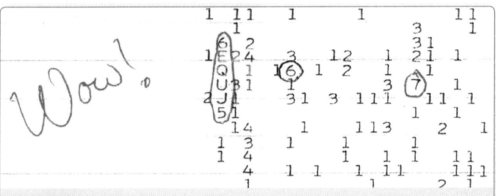

D *In 1977 researchers thought they had detected a simple signal from intelligent extraterrestrials, but it was never repeated.*

3 a Why can't scientists answer the question 'is there life beyond Earth?' at the moment?
b Why might they never be able to answer this question?

4 Explain why oxygen in the atmosphere of a planet might indicate that there is life there.

5 Scientists working on SETI projects have picked up many radio signals from space. Explain why they are not claiming to have discovered extraterrestrial life.

6 The *Phoenix* lander discovered frozen water beneath the soil on Mars during the Martian summer. Suggest why it is more likely that scientists will discover fossils of microbes on Mars rather than living microbes.

ResultsPlus
Watch Out!

The SETI projects are the only ones looking for *intelligent* life. The other methods would only find evidence that some kind of life may exist.

Skills spotlight

Many decisions about science and technology have to take into account economic pressures. Make some notes for a letter to your MP explaining why you think the government should (or should not) spend more money on looking for life in the Solar System.

Learning Outcomes

3.6 Describe the methods used to gather evidence for life beyond Earth, including space probes, soil experiments by landers, Search for Extraterrestrial Intelligence (SETI)

HSW 13 Describe the social, economic and environmental effects of decisions about the uses of science and technology

>>>>>>>>>>>>>>>>>>>>>>>>>>>>>>   How will the Sun destroy the Earth?   **235**

Life-cycles of stars

How are stars formed, and how do they change?

igure A shows an artist's impression of the Sun billions of years in the future. One day the Sun will expand enormously and may even swallow up the Earth. Long before that the extra heat will have killed all life.

A *The view from Earth in the future*

1 What 'fuel' is the Sun using to release energy?

2 How does a star form?

3 a What is a red giant?
b What is a white dwarf?

Star formation

A nebula is a cloud of dust and gases (mainly hydrogen). Stars form when a nebula is pulled together by gravity. As the contracting cloud gets more **dense**, it heats up and may begin to glow. As more and more mass is attracted the cloud's gravitational pull gets stronger and compresses the material even more. This is a **protostar**.

Eventually the temperatures and pressures in the centre of the protostar become high enough to force hydrogen **nuclei** to fuse together and form helium. **Fusion reactions** like this release a lot of energy as electromagnetic radiation. The outward pressure from the hot gases just balances the compression due to gravity. The star is now in the **main sequence** part of its life-cycle.

Life-cycles of stars like our Sun

Stars of similar sizes to our Sun remain **stable** for about 10 billion years, until they have fused most of their hydrogen into helium. When this happens the core of the star is not hot enough to withstand gravity, and it collapses. The outer layers will expand to form a **red giant** star. This is much larger than the original star.

B *The Eagle nebula, one of the places where new stars are forming in our galaxy. The dust and gas were lit up by the light of hot, young stars.*

Other fusion reactions happen inside red giants. The star will remain as a red giant for about a billion years before throwing off a shell of gas. The rest of the star will be pulled together by gravity and will collapse to form a **white dwarf** star. No fusion reactions happen inside a white dwarf and it will gradually cool over about a billion years to be a black dwarf.

4 Look at Figure A. Which stage of the Sun's life-cycle does it show?

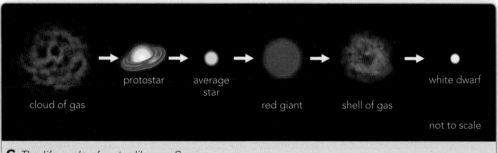

cloud of gas   protostar   average star   red giant   shell of gas   white dwarf

not to scale

C *The life-cycle of a star like our Sun*

Life-cycles of massive stars

Stars with considerably more mass than the Sun are hotter and brighter. They fuse hydrogen to helium faster and then become **red supergiants**. At the end of the red supergiant period the star will rapidly collapse and then explode. In this explosion, called a **supernova**, the outer layers of the supergiant are cast off and expand outwards.

If what is left is four or more times the mass of the Sun, gravity will pull the remains together to form a **black hole**. The gravitational pull of a black hole is so strong that not even light can escape from it. If the remains are not massive enough to form a black hole, gravity will pull them together to form a small, very dense star called a **neutron star**.

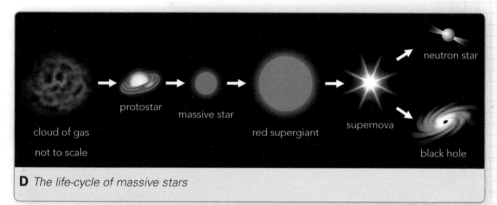

D *The life-cycle of massive stars*

E *Artist's impression of a black hole attracting gas from a nearby star. We can see the gas glowing as it is pulled into the black hole.*

Skills spotlight

Scientists develop theories to explain observations. However they cannot carry out experiments with stars. Suggest how astronomers worked out the theories that explain the star life-cycles shown in Figures C and D.

ResultsPlus
Watch Out!

Star life-cycles occur over *billions* of years, not *millions*.

H 5 What is a supernova?

H 6 Why won't the Sun form a black hole?

7 Describe two ways in which gravity has a part in the life-cycle of a star.

Learning Outcomes

3.11 Describe the evolution of stars of similar mass to the Sun through the following stages:
   a nebula
   b star (main sequence)
   c red giant
   d white dwarf

3.12 Describe the role of gravity in the life cycle of stars

H *3.13* Describe how the evolution of stars with a mass larger than the Sun is different, and may end in a black hole or neutron star

HSW 2 Describe the importance of creative thought in the development of hypothesis and theories

How do different theories explain the Universe?

Most cultures try to explain the world around them. Today, we find out about the world around us by scientific investigation. In the past, most cultures told stories describing a model of the Universe and how it works. Figure A shows an Ancient Egyptian model, with a sky goddess and an Earth god.

A

If a source of light is moving away from us, the light we detect has a longer wavelength than we would expect. Its light is 'shifted' towards the red end of the spectrum. This effect is called **red-shift**.

1 What does 'red-shift' tell us about distant galaxies?

The black lines can be seen in the light from the Sun.

Sun

distant galaxy

400   500   600   700

When a distant galaxy is moving away, the black lines in its spectrum shift towards the red end.

B *Red-shift*

In the early 20th century astronomers analysed the spectrum of light from our Sun and from distant galaxies. They could see lines or gaps in the spectrum. In the light from distant galaxies these lines were red-shifted, which showed that the galaxies are moving away from us. We now know that most galaxies are moving away from us. This can be explained if the Universe is expanding.

ResultsPlus
Watch Out!

The idea of the Big Bang is a *theory*, not a proven fact. It is supported by the evidence we have at the moment, but it is possible that the theory may have to be changed if new evidence is found.

Astronomers used this information and other data to work out theories that explained the origin and present state of the Universe. One of these came to be known as the **Big Bang theory**, first suggested in the 1930s. This says that the whole Universe and all the matter in it started out as a tiny point of concentrated energy about 13.5 billion years ago. The Universe expanded from this point and is still expanding. As the Universe expanded, gravity caused matter to clump together to form stars.

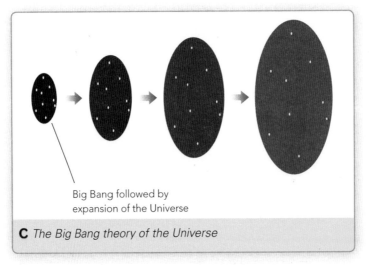

Big Bang followed by expansion of the Universe

C *The Big Bang theory of the Universe*

An alternative theory, the **Steady State theory**, was suggested in 1948. This theory says that the Universe has always existed and is expanding. New matter is continuously created within the Universe as it expands.

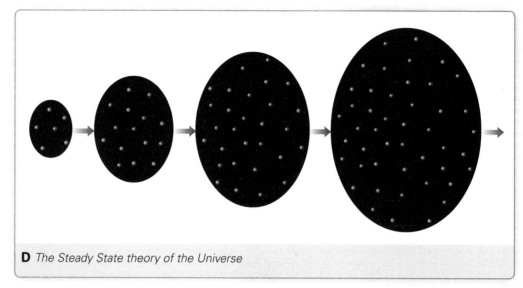

D *The Steady State theory of the Universe*

In 1964 two radio astronomers who were building a radio telescope detected microwave signals coming from all over the sky. At first they thought these were caused by a fault in their equipment, but eventually they realised that the signals were real. The astronomers realised that this was the radiation predicted by the Big Bang theory. The theory says that huge amounts of radiation were released at the beginning of the Universe and that this radiation should still be detectable as microwave radiation. It is called **cosmic microwave background (CMB) radiation**.

Both the Steady State and Big Bang theories say that the Universe is expanding. Red-shift in the light from other galaxies can be used as support for both theories. However, CMB radiation provides supporting evidence for the Big Bang theory. The Steady State theory cannot explain the CMB radiation. As there is more supporting evidence for it, the Big Bang theory is the one accepted by most astronomers today.

2 Which theory (or theories) says that:
a space is expanding
b new matter is being created all the time
c the Universe began about 13.5 billion years ago?

3 Explain why the movement of galaxies supports both the Steady State and Big Bang theories.

4 a What does CMB stand for?
b What is the CMB radiation?

5 Which theory (or theories) is supported by the following evidence:
a red-shift
b the cosmic microwave background radiation?

6 Describe how ideas about the beginning of the Universe have changed since 1900.

Skills spotlight

How are the scientific theories described on this page different from the explanation illustrated in Figure A?

Learning Outcomes

3.14 Demonstrate an understanding of the Steady State and Big Bang theories

3.15 Describe evidence supporting the Big Bang theory, limited to red-shift and cosmic microwave background (CMB) radiation

3.16 Recognise that as there is more evidence supporting the Big Bang theory than the Steady State theory, it is the currently accepted model for the origin of the Universe

HSW **1** Explain what scientific data is

HSW **2** Describe how data is used by scientists to provide evidence that increases our scientific understanding

How do we know that space is expanding?

Albert Einstein (1879–1955) came up with the Theory of General Relativity to show how matter, space, time and gravity were all linked. However, the theory predicted an expanding Universe, which he thought must be wrong. So he altered his theory! Edwin Hubble (1889–1953) showed that Einstein was right to begin with.

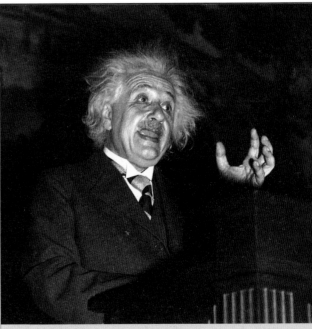

A *Einstein giving a lecture at the 8th American Scientific Conference in 1940*

1 a What is the difference in the frequency of the sound waves from an ambulance siren, between when it is approaching you and when it is heading away from you?

b How does the wavelength change?

H **c** Why is the wavelength different?

2 What is red-shift?

3 Draw a diagram to show how expanding space can cause a red-shift.

As an emergency vehicle's siren passes you, its **pitch** gets lower. This is the **Doppler effect**. The pitch depends on the frequency and wavelength of the **sound wave**. The sound waves behind a moving sound source become 'stretched', making the frequency lower and the wavelength longer and so the sound becomes lower in pitch. The opposite happens in front of the sound source. This is a similar effect to the one that causes red-shift.

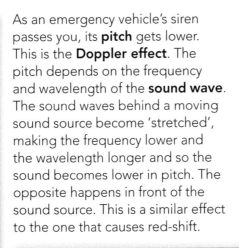

Stationary source. The wavelength is the same on both sides.

Moving source. The wavelength is longer behind the source, and shorter in front.

B *How the movement of a source affects the waves detected (viewed from above)*

Skills spotlight

H Scientists interpret data to provide evidence for developing theories. How did Hubble's interpretation cause Einstein to re-think his theory?

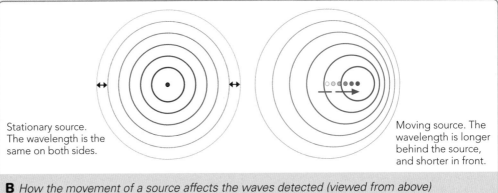

The visible spectrum of light from stars contains gaps. If these are red-shifted, the star is moving away from us. If they are shifted in the opposite direction (the wavelength and frequency are shorter than they should be) then the star is moving towards us.

In the 1920s, Hubble measured how far these gaps were shifted for many galaxies (see Figure C). He discovered that they were all red-shifted and concluded that all the galaxies were therefore moving away from us. The further away a galaxy is, the faster it is moving away from us. He interpreted this to mean that the Universe itself must be expanding.

Evidence for theories

Astronomers used the Big Bang theory to predict that radiation from the young, hot, early Universe would have spread out. The radiation would originally have had high energy and very short wavelengths. The expansion of the Universe, as the waves travelled through it, made the wavelength increase.

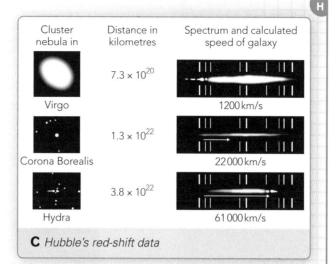

Cluster nebula in	Distance in kilometres	Spectrum and calculated speed of galaxy
Virgo	7.3×10^{20}	1200 km/s
Corona Borealis	1.3×10^{22}	22000 km/s
Hydra	3.8×10^{22}	61000 km/s

C Hubble's red-shift data

When cosmic microwave background radiation was discovered in 1965 it matched the prediction made by the Big Bang theory.

D A map of CMB radiation made by the WMAP satellite. The colours show tiny variations from the average microwave wavelength of around 1 mm.

Both the Big Bang and Steady State theories predict that the Universe is expanding and are supported by the evidence from red-shift. However, the Steady State theory does not predict CMB radiation or the abundances of hydrogen and helium. There is much more supporting evidence for the Big Bang theory, so this is now the accepted model.

ResultsPlus
Watch Out!

Red-shift means that light from a distant galaxy has a longer wavelength (it is shifted towards the red end of the spectrum). It does not mean that the light from the galaxy is red.

H 4 Explain why the red-shift of galaxies shows that the universe is expanding.

H 5 The Sunflower Galaxy is 3.5×10^{20} km away and the Southern Pinwheel Galaxy is 1.4×10^{19} km away. Which galaxy is moving away from us more quickly? Explain your answer.

H 6 Explain why the Big Bang theory is now the accepted model of the Universe.

H 7 Explain how the relationship between the red-shift of a galaxy and its distance away from us can be used as a way of measuring the distance to newly discovered galaxies.

Learning Outcomes

3.17 Describe that if a wave source is moving relative to an observer there will be a change in the observed frequency and wavelength

H 3.18 Demonstrate an understanding that if a wave source is moving relative to an observer there will be a change in the observed frequency and wavelength

H 3.19 Describe the red-shift in light received from galaxies at different distances away from the Earth

H 3.20 Explain why the red-shift of galaxies provides evidence for the Universe expanding

H 3.21 Explain how both the Big Bang and Steady State theories of the origin of the Universe both account for red-shift of galaxies

H 3.22 Explain how the discovery of the CMB radiation led to the Big Bang theory becoming the currently accepted model

HSW 3 Describe how phenomena are explained using scientific models

How do we use sound waves that are too low for us to hear?

Just as there are electromagnetic waves that our eyes cannot detect, there are sound waves that our ears cannot detect. However, some scientists think that our bodies can detect these infrasound waves in some way and this may cause some people to think they have seen a ghost.

A *Ghosts in photos cannot be explained by infrasound waves are likely to have everyday explanations.*

Sound and light both travel as waves, and the waves can be described in similar ways. The **frequency** of a wave is the number of complete waves that pass a point in one second and is measured in **hertz (Hz)**. All sound waves are **longitudinal** vibrations that travel through a **medium** (a material). The medium can be a solid, a liquid or a gas. The frequency of a sound wave determines the **pitch** of the sound (how high or low it is).

> 1 a What does 'frequency' mean?
> b What are the units for frequency?
>
> 2 A vibration has a frequency of 200 Hz. Is it an infrasound? Explain your answer.
>
> 3 Why do you think infrasound is more useful for communication to animals like elephants than sound?

Humans can hear sounds in the frequency range 20 to 20 000 Hz (or 20 kHz). Sound waves with low frequencies produce sounds with a low pitch. Sounds with frequencies below 20 Hz are called **infrasound**. We cannot hear infrasounds, but we can detect them using microphones.

Some animals can produce and hear infrasound waves. Elephants, giraffes and some other animals communicate with each other using infrasound. Infrasound waves travel further in air than sound waves of higher frequencies before they become too faint to detect. Whales also communicate using infrasound, and their calls can travel for hundreds of kilometres through the ocean.

Using infrasound

Infrasound is useful for studying animals. Animals such as elephants or giraffes can move many kilometres each day and often live in forests where it is difficult to find them. Biologists can use microphones that detect infrasounds to find the animals they are studying and to find out how they move around throughout the year.

B *Elephants were once thought to have mind-reading abilities because one herd seemed to know what was happening to another herd many kilometres away.*

Natural events such as volcanic eruptions create infrasound waves as well as sounds that we can hear. Many volcanoes are in remote places, making it difficult for scientists who study volcanoes to know when one is erupting. Infrasound waves produced by the eruption can be detected by sensors a long way from the volcano. The eruption can then be continuously monitored.

Meteors are rocks that fall into the atmosphere from space. Some meteors burn up in the atmosphere and some explode. However some survive and hit the ground (when they are called meteorites). A large meteorite hitting a populated area could cause a major disaster.

C *An infrasound detector in Greenland*

We sometimes see meteors as streaks of light in the sky. However meteors that enter the atmosphere over unpopulated areas may not be seen. Scientists can use infrasounds to detect the passage of meteors through the atmosphere and also detect any that explode. This helps scientists to work out how many meteors enter the Earth's atmosphere and what the risks of meteor impacts may be.

D *In 1908 this forest in Tunguska (in Russia) was flattened over hundreds of square kilometres. Scientists think a huge meteor exploded in the air with the energy of a large nuclear bomb.*

ResultsPlus
Watch Out!

Sound waves are longitudinal, but light waves are transverse. Frequency and wavelengh are used to describe both types of waves (see P1.5).

4 Explain why infrasound is useful for detecting animals in forests.

5 Why do you think scientists don't just rely on people reporting meteor explosions to them but also use infrasound detectors?

6 Explain why infrasound can sometimes be more useful than satellite photographs for monitoring volcanoes.

Learning Outcomes

4.4 Recall that sound with frequencies less than 20 hertz, Hz, is known as infrasound

4.5 Describe uses of infrasound, including: **a** communication between animals **b** detection of animal movement in remote locations **c** the detection of volcanic eruptions and meteors

1.15 Use of both the equations below for all waves:

wave speed (metre/second, m/s) = frequency (hertz, Hz) × wavelength (metre, m) $v = f \times \lambda$

wave speed (metre/second, m/s) = distance (metre, m)/time (second, s) $v = \dfrac{x}{t}$

HSW 12 Describe the benefits, drawbacks and risks of using new scientific and technological developments

How do we use sound waves that are too high for us to hear?

Many zoos try to help endangered species by breeding animals. Some animals are very difficult to breed in captivity, and zoo vets need to be able to monitor them carefully when they are pregnant. Ultrasound scans can be used to check how the baby animals are developing.

1 What is ultrasound?

2 Why do biologists need to use special equipment to detect the noises that bats make?

Humans can detect sound waves in the frequency range 20 to 20 000 Hz (or 20 kHz). Sounds made by waves with higher frequencies than this are called **ultrasound**. Some animals, such as dolphins, use ultrasound to communicate with each other.

Sonar

Some animals, such as bats, can detect obstacles and other objects around them using ultrasound waves. The ultrasound waves made by the animals are **reflected** by things around them, and the animals listen for the echoes.

Humans can use a similar method to detect things that we cannot see. **Sonar** equipment carried on ships or submarines can be used to find the depth of the sea or to detect fish.

Skills spotlight

Bats can fly around and catch insects in the dark without bumping into things. Two ways of explaining this observation could be if the bats had very sensitive eyes that could detect even tiny amounts of light or if they used a type of sonar. What data could scientists collect to test these two hypotheses?

3 Why is ultrasound used for investigating the sea floor rather than light?

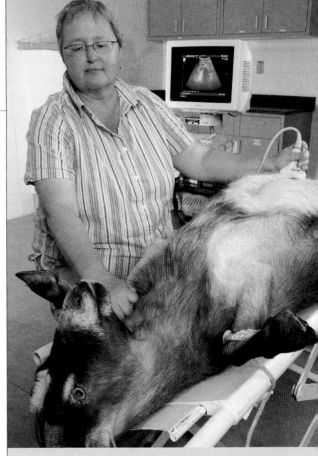

A Ultrasound being performed on a pregnant goat as part of study on crooked calf syndrome

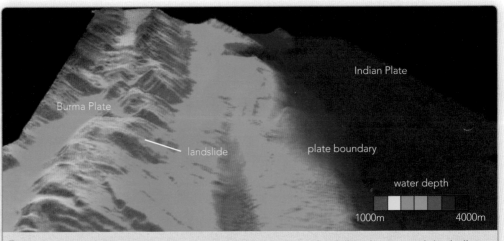

Indian Plate

Burma Plate

landslide

plate boundary

water depth

1000m   4000m

B This sonar image of the sea floor shows the underwater landslide that caused the Indian Ocean tsunami in 2004. The tsunami killed over 200 000 people.

A loudspeaker on the ship emits a pulse of ultrasound. This spreads out through the water, and some of it is reflected by the sea bed. A special microphone on the ship detects the echo, and the sonar equipment measures the time between the sound being sent out and the echo returning. The distance travelled by the sound wave can be worked out using this equation:

distance (metre, m) = speed (metre/second, m/s) × time (second, s)

Ultrasound scans

Ultrasound can also be used to make images of things inside the body. One common use is to make detailed images of unborn babies so that doctors can monitor how well the foetus is developing. When an **ultrasound scan** is made, some of the ultrasound waves are reflected each time the waves pass into a different medium.

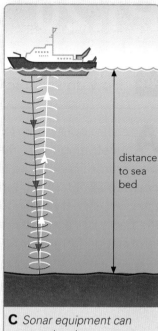

distance to sea bed

C *Sonar equipment can map the shape of the sea bed.*

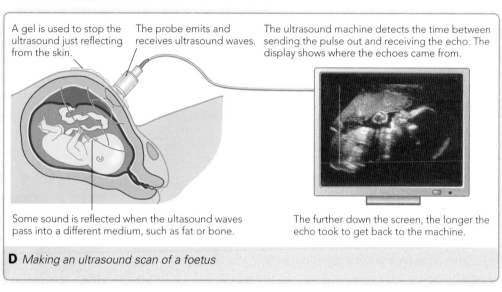

A gel is used to stop the ultrasound just reflecting from the skin.

The probe emits and receives ultrasound waves.

The ultrasound machine detects the time between sending the pulse out and receiving the echo. The display shows where the echoes came from.

Some sound is reflected when the ultasound waves pass into a different medium, such as fat or bone.

The further down the screen, the longer the echo took to get back to the machine.

D *Making an ultrasound scan of a foetus*

4 A ship detects a sonar pulse 3 seconds after it was emitted. The speed of sound in the sea water is 1500 m/s.
a How far has the sound travelled? b How deep is the water?

5 Suggest why ultrasound is used in medical scans rather than:
a visible light b X-rays.

6 Explain how an ultrasound scanner works.

7 Explain how sonar works and suggest how it could be used to detect fish.

Learning Outcomes

4.1 Recall that sound with frequencies greater than 20 000 hertz (Hz) is known as ultrasound

4.2 Describe uses of ultrasound, including: a sonar b communication between animals c foetal scanning

4.3 Calculate depth or distance from time and velocity of ultrasound

1.15 Use of both the equations below for all waves:

wave speed (metre/second, m/s) = frequency (hertz, Hz) × wavelength (metre, m) $v = f \times \lambda$

wave speed (metre/second, m/s) = distance (metre, m)/time (second, s) $v = \dfrac{x}{t}$

HSW 5 Plan to test a scientific idea, answer a scientific question, or solve a scientific problem by selecting appropriate data to test a hypothesis

How do seismic waves travel through the Earth?

Archaeologists can detect buried objects using ground-penetrating radar. Reflections of the electromagnetic waves produced by the radar can be used to detect objects down to about 15 m. Geologists can investigate rocks down to about 2 km by drilling deep holes. To find out what the Earth is like at greater depths, geophysicists can use information from seismic waves.

A *The machine sends pulses of microwaves into the Earth and detects any waves reflected by buried objects.*

1 What is:
a a seismic wave
b a seismometer?

2 Write down two things that can cause seismic waves.

3 Write down two differences between P waves and S waves.

Movements inside the Earth, such as **earthquakes**, cause waves to be transmitted through the Earth. These are called **seismic waves**. When the waves reach the surface of the Earth, the ground shakes. **Seismometers** are instruments that can detect these waves.

The place inside the Earth where rock suddenly moves or fractures is called the **focus** of an earthquake. The point on the surface of the Earth directly above the focus is called the **epicentre**. An earthquake causes two types of wave within the Earth. **P waves** are longitudinal waves and **S waves** are transverse waves. There are also some waves that travel along the surface of the Earth. P waves travel faster than S waves.

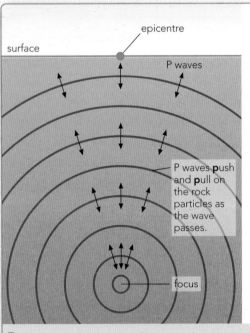

surface
epicentre
P waves

P waves push and pull on the rock particles as the wave passes.

focus

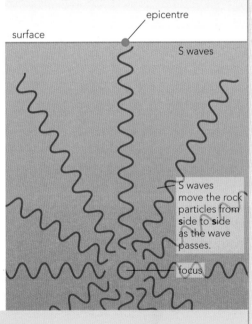

surface
epicentre
S waves

S waves move the rock particles from side to side as the wave passes.

focus

B *Seismic waves caused by an earthquake*

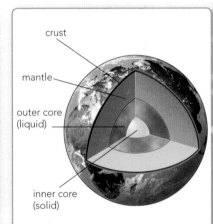

crust
mantle
outer core (liquid)
inner core (solid)

C *The surface of the Earth is called the **crust**. Layers inside the Earth are the **mantle** and the **core**. Seismic waves are reflected and refracted at boundaries between the different layers.*

Investigating the Earth

Scientists investigate the layers of rocks near the surface of the Earth by setting off small explosions or even dropping a large mass from a truck. They place seismometers to record the seismic waves at different distances from the source. The seismic waves are reflected and **refracted** when they pass into different rocks. The travel times of these waves give information about the rocks beneath the surface. Oil companies use these methods to search for new oil and gas fields.

To investigate deeper structures within the Earth, scientists rely on the much greater amount of energy released by earthquakes. S and P waves are reflected and refracted at the boundaries between the Earth's crust, mantle and core.

D *Geologists conducting a seismic survey*

The properties of rocks gradually change with depth, causing the speed of seismic waves to increase gradually. This makes the waves gradually refract upwards and travel on a curved path. When there is a sudden change in rock properties, such as a layer of a different kind of rock, the waves are refracted sharply at this boundary and some of the waves may be reflected.

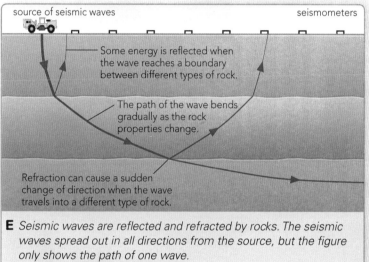

source of seismic waves — seismometers

Some energy is reflected when the wave reaches a boundary between different types of rock.

The path of the wave bends gradually as the rock properties change.

Refraction can cause a sudden change of direction when the wave travels into a different type of rock.

E *Seismic waves are reflected and refracted by rocks. The seismic waves spread out in all directions from the source, but the figure only shows the path of one wave.*

Scientists use information from seismometers all around the world to analyse the seismic waves produced by earthquakes. Analysis of this data can help them to model the structure of the Earth's interior.

Skills spotlight

Suggest the benefits and drawbacks of using the following methods to produce seismic waves: trucks such as the one shown in Figure D, explosions, nuclear explosions, earthquakes.

ResultsPlus
Watch Out!

There are *transverse* (S) waves and *longitudinal* (P) waves. Waves can be *reflected* or *refracted*. Make sure you understand these terms.

4 What two processes can affect the path of seismic waves?

H 5 a Why do seismic waves follow curved paths within the Earth?
b What can cause a sudden change in the direction of travel of a seismic wave?

6 Explain how scientists can use seismic waves to work out the structure of the rocks beneath the surface.

Learning Outcomes

4.6 Recall that seismic waves are generated by earthquakes or explosions

4.9 Recall that seismic waves can be longitudinal (P) waves and transverse (S) waves and that they can be reflected and refracted at boundaries between the crust, mantle and core

H 4.11 Demonstrate an understanding of how P and S waves travel inside the Earth including reflection and refraction

HSW 12 Describe the benefits, drawbacks and risks of using new scientific and technological developments

P1.21 Earthquakes

⠿ **Why can't scientists predict when earthquakes will happen?**

Hundreds of people die in earthquakes every year, and if there is a big earthquake, hundreds of thousands of people can die. Some deaths are caused by people being trapped in collapsed buildings, and others occur due to disease or starvation if the earthquake destroys water supplies or sewage systems, or prevents food being delivered. So why can't scientists predict where and when earthquakes will happen so that governments can evacuate people?

A *The Dominican Defensa Civil team searching for survivors after a massive earthquake in Haiti in January 2010.*

The outermost layer of the Earth is made of **plates**. These plates all have forces on them which make them move relative to each other. The movement does not happen smoothly, because friction between the edges of the plates stops them moving.

The forces on the plates build up until they are big enough to overcome the friction, and then the plates move with a sudden jerk. Each jerk causes an earthquake. Some earthquakes start near the surface of the Earth, and some start deep underground.

Diagram B shows a model that can be used to investigate earthquakes. The bungee cord represents the forces trying to move the plates. An 'earthquake' happens when the brick moves.

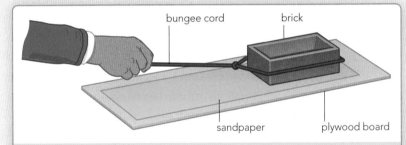

B *An earthquake model. This model can be used to investigate how often earthquakes occur as the forces build up. The bungee cord represents the forces trying to move the plates. An 'earthquake' happens when the brick moves.*

Your task

You are going to plan an investigation that will allow you to find out about earthquakes and the forces acting in the Earth's crust. The hypothesis is that the deeper the location of an earthquake, the more force is needed to start the plates moving, so there should be a longer interval between earthquakes. Your teacher will provide you with some materials to help you organise this task.

Learning ⊙utcomes

4.7 Investigate the unpredictability of earthquakes, through sliding blocks and weights

Build Better Answers

When completing an investigation like this, one of the skills you will be assessed on is your ability to *process the data*. There are 4 marks available for this skill. Here are two student extracts that focus on this skill. Other skills that you need for the assessment are dealt with in other lessons.

Student extract 1 — A basic response for this skill

This is not how to work out the mean – make sure that you use the right mathematical process.

> I have taken all the results in my results table and worked out a mean for each set of data. To do this I have put all my data in order and then have taken the middle piece of data. I have then tried to plot this on a scatter graph.

Primary and secondary evidence should be processed if possible.

A scatter graph shows you if there is a link between two variables – it is the right way of presenting the evidence.

Student extract 2 — A good response for this skill

This is a good way of showing all your data in one place.

> I have taken the data from each set of results and worked out a mean in each case. I have then done the same for my secondary evidence. I have added a column to my results table to show this information. I have then plotted the mean against the mass in a line graph. I have made sure that the graph is plotted correctly with correct units and headings.

Line graphs should only be used when you have a set of results which are samples from a continuous experiment.

ResultsPlus

To access 2 marks
- Attempt to process all your collected evidence
- Use maths skills if appropriate
- Present the processed evidence in an appropriate way

To access 4 marks
- Process all your evidence in an appropriate way, using maths skills if appropriate
- Present it in a way that allows you to draw a conclusion

How can seismic waves tell us where an earthquake happened?

The Indian Ocean tsunami of 2004 struck without warning on Boxing Day and killed hundreds of thousands of people. A warning system is now set up in the Indian Ocean to detect earthquakes that may cause tsunamis so that people can be warned to leave areas at risk.

A *The tsunami approaching*

Detecting earthquakes

Scientists use a network of seismometers around the world to detect earthquakes and work out where they occurred. If an earthquake happens in a remote region, scientists can often work out its location and alert emergency services to get people out of the disaster area.

> **1** Which type of waves from an earthquake arrives first at a seismometer?

The seismometer trace in Figure B shows waves from an earthquake. The P waves arrive first because they travel faster than S waves. The time difference between the arrival of the first P waves and the first S waves can be used to work out how far away from a seismometer an earthquake occurred. Data from three or more seismometers can be used to work out the location of the epicentre, or point where the earthquake originated.

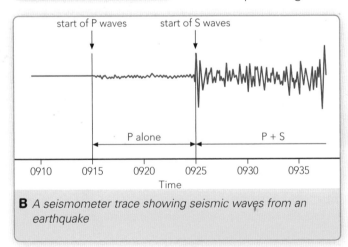

B *A seismometer trace showing seismic waves from an earthquake*

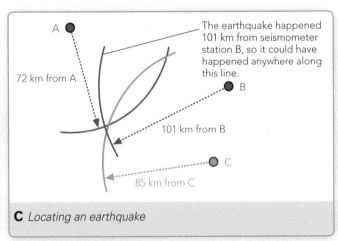

C *Locating an earthquake*

Predicting earthquakes and tsunamis

Figure D shows the locations of large earthquakes around the world. The outermost layer of the Earth is made of **tectonic plates**, pushed by very slow moving **convection currents** in the mantle. These plates are all moving relative to each other. Friction between the plates stops them moving until the forces are big enough to overcome the friction and then they move with a sudden jerk. Each jerk causes an earthquake.

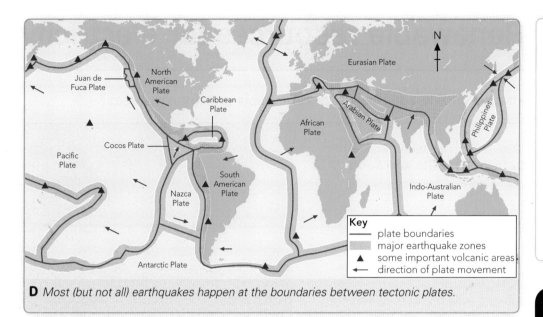

D *Most (but not all) earthquakes happen at the boundaries between tectonic plates.*

2 a Give two reasons why it is useful to be able to locate where an earthquake occurred.
b Explain how scientists do this.

3 Look at Figure C. Explain why you cannot just use two seismometers to find the location of an earthquake.

ResultsPlus
Watch Out!

The rock mantle below the tectonic plates is solid. The convection currents happen extremely slowly in this solid rock.

Scientists can use the plate boundaries to predict where earthquakes are likely to happen. However, it is not currently possible to measure the forces trying to move the plates or the friction between them. This makes it difficult to predict when a sudden movement will happen.

If the earthquake happens under the sea, the movement of the sea floor may cause a huge wave called a **tsunami** (pronounced '*soo-na-me*'). However, scientists cannot tell whether or not a tsunami will happen from seismometer traces. Tsunami warning systems include pressure sensors that detect the tsunami waves in water. Tsunamis travel at around 220 m/s in deep water, so people on the far side of an ocean from a tsunami may get several hours' warning.

Skills spotlight

There are some scientific questions that cannot currently be answered. Write a short magazine article from a geologist explaining why she can predict *where* earthquakes may happen but not *when*.

4 What causes earthquakes?

5 Children in California, on the west coast of the USA, have to practise 'earthquake drills'. Suggest why this is not necessary in the UK.

6 The coast of Sri Lanka is about 1000 km from the epicentre of the earthquake that caused the Indian Ocean tsunami. How much warning would the people have had if there had been a warning system in place?

7 A tsunami warning system that only used seismometers could produce a lot of false alarms. Suggest why and explain why this kind of system might not be very helpful.

Learning Outcomes

4.8 Explain why scientists find it difficult to predict earthquakes and tsunami waves even with available data

4.10 Explain how data from seismometers can be used to identify the location of an earthquake

4.12 Explain how the Earth's outermost layer, composed of (tectonic) plates, is in relative motion due to convection currents in the mantle

4.13 Demonstrate an understanding of how, at plate boundaries, plates may slide past each other, sometimes causing earthquakes

HSW 4 Identify questions that science cannot currently answer, and explain why these questions cannot be answered

What are the advantages and disadvantages of renewable energy resources?

This is a proposal for a 'solar tower' in Australia. It would be taller than the highest building in the world and would generate electricity using energy from the Sun.

A

Electricity is a flow of charged particles. The **current** is the rate of charge flowing. The **voltage** is an 'electrical pressure' which gives a measure of the amount of energy transferred by the charges to an electric component. We use electricity for many things, and it all has to be generated using other forms of energy.

Renewable energy resources are resources that will not run out. Most renewable energy resources do not cause pollution or greenhouse gases when used to generate electricity because no fuel is burned.

1 What advantage do almost all renewable energy resources have over fossil fuels for generating electricity?

Solar cells convert **solar energy** directly into electrical energy. In the type of solar power station shown in Figure A, solar energy is used to heat the air under the glass. The hot air rises up the tower, turning turbines as it moves. Another type of solar power station is shown in Figure B. Solar energy is not available all the time.

Hydroelectricity is generated by falling water in places where water can be trapped in high reservoirs. It is available at any time (as long as the reservoir does not dry up). A hydroelectric power station can also be started and stopped very quickly.

Wind turbines can be used to generate electricity (as long as the wind speed is not too slow or too fast). A lot of wind turbines are needed to produce the same amount of energy as a fossil-fuelled power station and some people think they spoil the landscape.

Geothermal energy is heat transferred from hot underground rocks. In some places the rocks are hot enough to turn underground water into steam, which can turn turbines in power stations to generate electricity. Some gases dissolved in the steam can cause pollution.

B A solar power station. Sunlight is focused on the tower in the middle by the mirrors and the energy absorbed turns water into steam.

C *A geothermal power station in Iceland. Surplus heated water provides a bathing lake.*

D *A possible design for a turbine to generate electricity from tidal currents*

Tidal power can generate electricity when turbines in a huge barrage (dam) across a river estuary turn as the tides flow in and out. Tidal power is not available all the time but is available at predictable times. There are not many places in the UK that are suitable for barrages and they may affect birds and other wildlife that live or feed on tidal mudflats. Underwater turbines can also be used to generate electricity, as seen in Figure D.

Wave power can generate electricity when floating electrical generators move up and down. In coastal power stations, ocean waves can force air up pipes in the power station, and the moving air turns turbines to generate electricity.

> 2 Describe two different ways in which electricity can be produced using:
> a solar energy
> b tidal power.
>
> 3 Explain which renewable resources:
> a are available all the time
> b are available at predictable times
> c can only be used in certain places
> d depend on the weather.
>
> 4 Suggest which renewable energy resources:
> a cannot be used in the UK
> b would be more useful in summer
> c would be more useful in winter.
>
> 5 Explain why it would be difficult to generate all our electricity from renewable resources. Give examples to illustrate your answer.

Skills spotlight

Decisions on scientific issues often need to consider many factors. What factors would a local council need to consider when choosing renewable energy resources for a community?

Learning Outcomes

5.1 Describe current as the rate of flow of charge and voltage as an electrical pressure giving a measure of the energy transferred

5.5 Discuss the advantages and disadvantages of methods of large-scale electricity production using a variety of renewable resources

HSW **13** Describe the social, economic and environmental effects of decisions about the uses of science and technology

Why did radioactive dust fall across Europe in the 1980s?

What are the advantages and disadvantages of non-renewable energy resources?

France depends mainly on nuclear power for electricity generation, but nuclear power is banned under Austrian law. Austria finished building a nuclear power plant in 1978, but in a referendum, the Austrian people voted against turning it on. In 2009, a photovoltaic plant was put into operation on part of the plant.

A The Zwentendorf nuclear power station in Austria cost nea £1 billion to build but was never turned on.

ResultsPlus
Watch Out!

Nuclear power is non-renewable because supplies of nuclear fuel will eventually run out. in the same way that fossil fuels will eventually run out.

Most of the electricity used in the UK is generated using **non-renewable resources**. Non-renewable resources will eventually run out. They include **fossil fuels** such as coal, oil and natural gas. **Nuclear power** stations use radioactive metals such as uranium or plutonium as their energy source. These metals will also run out one day.

Fossil-fuelled power stations produce waste gases, which can cause problems. For example, carbon dioxide emissions contribute to **climate change** and gases such as sulfur dioxide and nitrogen oxides can cause **acid rain**. Coal-fired power stations also produce dust and ash. There are various ways of reducing this pollution, but these cost money.

Not all fossil fuels cause the same amount of pollution. Natural gas is 'cleaner' than coal because it contains less nitrogen and does not contain sulfur. Natural gas power stations also emit less carbon dioxide than other power stations producing the same amount of electricity.

1 List two disadvantages of fossil-fuelled power stations compared to renewable ways of generating electricity.

2 Describe two advantages of using natural gas instead of coal to generate electricity.

Fuel	Estimated time left
coal	164 years
gas	67 years
oil	41 years
uranium	265 years

B How long our reserves of fossil fuels will last at the rate we are currently using them.
Source: World Coal Institute/IEA

C A fossil-fuelled power station

Nuclear power stations do not emit any carbon dioxide or other gases. However, the waste they produce is radioactive, and some of it will stay radioactive for millions of years. This waste must be sealed into concrete or glass and buried safely so the radioactivity cannot damage the environment. A nuclear power station also needs to be carefully **decommissioned** (dismantled safely) at the end of its life so that no radioactive materials escape into the environment. It costs a lot more to build and to decommission a nuclear power station than a fossil-fuelled one.

There are not many accidents in nuclear power stations, and the stations are designed to contain any radioactive leaks. However, if a major accident occurs it can have very serious consequences.

In spite of the possible pollution they cause, fossil-fuelled and nuclear power stations have some important advantages.

- At the moment there is a good supply of both fossil fuels and nuclear fuel.
- They produce cheaper electricity than electricity from renewable resources.
- They do not depend on the weather or the tides, so electricity is available all the time.

D At 01.23 on 26 April 1986, a nuclear reactor at Chernobyl, Ukraine, exploded. Winds blew radioactive dust across Europe and over 500 000 soldiers were called in to make the area safe. One report estimates that about 140 000 people will die as a result of the accident.

3 a Write down one advantage of nuclear power over fossil-fuelled power stations.
b Write down two disadvantages.

4 a Why do you think a nuclear power station costs more to build?
b Why is decommissioning it properly very important?

5 How can a nuclear accident affect people in other parts of the world?

6 Describe two advantages and two disadvantages of a fossil-fuelled power station compared to a solar power station.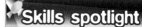

Skills spotlight

Technological developments can have benefits, drawbacks and risks. Develop an argument in favour or against building any more nuclear power stations in the UK.

Learning Outcomes

5.5 Discuss the advantages and disadvantages of methods of large-scale electricity production using a variety of non-renewable resources

HSW *12* Describe the benefits. drawbacks and risks of using new scientific and technological developments

HSW *13* Describe the social, economic and environmental effects of decisions about the use of science and technology

>>>>>>>>>>>>>>>>>>>>>>>>>>>

How can you make electricity?

P1.25 Investigating generators

What are the factors that affect the amount of electricity produced by a generator?

Figure A shows the world's smallest wind-up torch. It has a rechargeable cell inside. Turning the handle operates a tiny generator that produces electricity to recharge the cell.

Maths skills

You are testing the hypothesis that the speed of the magnet is directly proportional to the voltage. Two quantities are said to be in direct proportion if they increase and decrease in the same ratio. A directly proportional relationship is indicated by a graph where the line is a straight line passing through the origin.

Inverse proportion is when one quantity increases at the same rate as the other quantity decreases.

If you move a magnet into a coil of wire, a voltage is induced in the wire. If the wire is part of a complete circuit, the voltage will cause a current to flow in the circuit. Larger voltages produce larger currents. This process is called electromagnetic induction, and the current is called an induced current. You get the same effect if you move the coil instead of the magnet. A current is produced whenever a coil of wire and a magnet move relative to one another.

A A wind-up torch

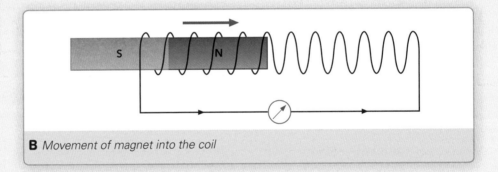

B Movement of magnet into the coil

Your task

You are going to plan an investigation that will allow you to find out what happens when a magnet passes through a coil of wire at different speeds and about the current induced. You are going to test the hypothesis that the size of the induced current increases as the speed of the magnet increases. Your teacher will provide you with some materials to help you organise this task.

Learning Outcomes

5.7 Investigate factors affecting the generation of electric current by induction

ResultsPlus
Build Better Answers

When completing an investigation like this, one of the skills you will be assessed on is your ability to *evaluate how strong your conclusion is.* There are 4 marks available for this skill. Here are two student extracts that focus on this skill. Other skills that you need for the assessment are dealt with in other lessons.

Student extract 1 | A basic response for this skill

Use both primary and secondary evidence.

You should also comment on any way that you can make your conclusion even stronger.

> I think that my conclusion is a strong one because my results showed a definite pattern and there were no anomalous results. The information I got from the textbook also showed the same pattern and made me come to the same conclusion. I could have made my conclusion stronger by taking more repeat results which would have given me more data to work with and lessened the effect of any anomalous results.

Say if you think your conclusion is strong or weak.

Then say why you have this opinion.

Student extract 2 | A good response for this skill

Make sure you explain what kind of experiments. Do you mean taking repeat readings or extending the practical in some way?

> I think that my conclusion is a strong one. I think that I could make it stronger by doing more experiments. I think it is good because it follows a pattern and follows the pattern in the textbook and there are no anomalous results. I think that my conclusion also matched my scientific knowledge which is that the faster the magnet moves through the coil the larger the induced current will be.

This extract covers more points than the one above but it is not very well organsed and this could cost you marks if it is not obvious that you have covered everything.

You need to refer to your scientific knowledge and explain if your conclusion matches your scientific understanding.

ResultsPlus

To access 2 marks
- Evaluate how well all your evidence supports your conclusion
- Suggest how your evidence can be improved to strengthen your conclusion

To access 4 marks
You also need to:
- Evaluate how well relevant scientific ideas support your conclusion
- Suggest how the investigation could be extended to support your conclusion

P1.26 Generating electricity

How is electricity generated?

The green squares in the pavement convert movement into electricity when you step on them. Some of this electricity is used to make the squares glow. The rest could be used to power street lights or signs. New technologies like this may help us to reduce the electricity used to light up streets and buildings but they will not be able to produce all the electricity we need, so we will still need big power stations to generate electricity.

1 Write down *three* ways in which you can *reduce* the size of an induced current.

If you move a piece of wire in a **magnetic field**, an electric current will flow in the wire. This process is called **electromagnetic induction**, and a current produced in this way is an **induced current**.

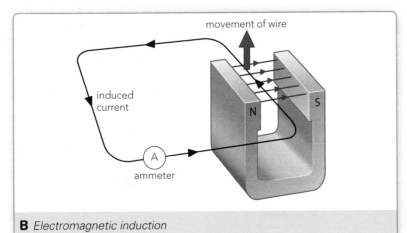

B *Electromagnetic induction*

The size of the induced current can be increased by:

- using a coil of wire, or putting more turns on the coil
- using an iron core inside the coil of wire
- using stronger magnets
- moving the wire faster.

The direction of the current in Figure B can be changed by changing the direction:

- of movement of the wire
- of the magnetic field.

2 A bicycle has lights that are powered by a dynamo.
a Would the lights be brighter when going uphill or downhill?
b Explain your answer.

3 Write down *two* advantages and *two* disadvantages of using a dynamo to provide the electricity to power bicycle lights.

To create a continuous induced current, you must keep the magnet moving *relative* to the coil of wire. This means that it does not matter whether it is the coil or the magnet that moves.

Some bicycles have **dynamos** to produce electricity for their lights. A magnet spins inside a coil of wire which induces a current. The current produced by a simple dynamo is always flowing in the same direction and is called **direct current** (DC). Wind-up radios and torches work in a similar way.

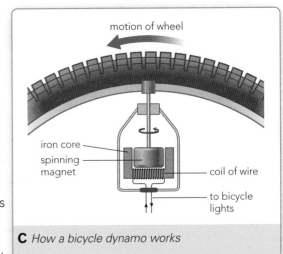

C *How a bicycle dynamo works*

Figure D shows a **generator**, similar to one you could use in a school laboratory. Electricity induced in the coil of the generator is transferred to a circuit through the **slip rings**, which touch **carbon brushes** attached to the circuit. The slip rings and brushes allow electricity to flow between the moving coil and the stationary circuit without wires becoming twisted.

Since each side of the coil goes up through the magnetic field and then down again, the direction of the induced current changes. This happens many times each second. A current that changes direction like this is called **alternating current** (AC).

The generators in power stations need to induce a very high voltage. Strong permanent magnets are very expensive, so generators often use **electromagnets** instead. Electromagnets are magnets that are created using electricity. They can create much more powerful magnetic fields than permanent magnets.

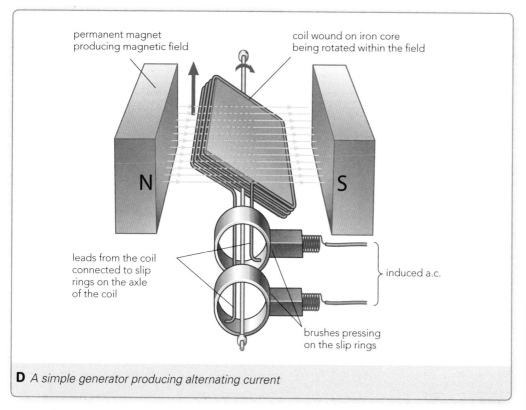

permanent magnet producing magnetic field

coil wound on iron core being rotated within the field

N

S

induced a.c.

leads from the coil connected to slip rings on the axle of the coil

brushes pressing on the slip rings

D *A simple generator producing alternating current*

7 Explain why generators produce alternating current.

4 Explain the difference between direct and alternating current.

5 Why are electromagnets used in power stations instead of permanent magnets?

6 Look at Figure B. What will happen to the direction of the current if you move the wire downwards *and* swap the poles of the magnet? Explain your answer.

ResultsPlus
Watch Out!

The wire must move across the magnetic field to induce a current. If the wire in Figure B moves from left to right (or right to left) there will not be an induced current.

Learning Outcomes

5.6 Demonstrate an understanding of the factors that affect the size and direction of the induced current

5.8 Explain how to produce an electric current by the relative movement of a magnet and a coil of wire
a on a small scale
b in the large-scale generation of electrical energy

5.9 Recall that generators supply current which alternates in direction

5.10 Explain the difference between direct and alternating current

HSW **12** Describe the benefits, drawbacks and risks of using new scientific and technological developments

How is electricity sent around the country?

The electricity used by electric trains can be sent through 'live' rails or through overhead lines. The voltages used are high enough for sparks to jump through the air, so you can be electrocuted just by standing close to overhead lines or a live rail. So why do they use such dangerously high voltages?

A *Electricity jumping between the live rail and the train*

Skills spotlight

The use of technology can have benefits, drawbacks and risks. What are some of the advantages and disadvantages of using transformers to increase the voltage of electricity sent through the National Grid?

Electricity is sent from power stations to homes, schools and factories by a system of wires and cables called the **National Grid**. Electricity passing through a wire will heat it up. The heat produced in the transmission lines in the National Grid is wasted energy. If the voltage of the electricity passing through the wires is increased, less energy is wasted as heat and the **efficiency** is improved. Power stations produce alternating currents at 25 kV (25 000 V). This is changed to 400 kV before the electricity is sent around the country.

1 Write these National Grid voltages in the order that they occur in Figure B, starting with the voltage at the power station: 11 kV, 25 000 V, 33 kV, 0.23 kV, 400 kV.

2 Explain why the voltage is increased before electricity is sent through the National Grid.

3 Look at Figure B. The transformers are labelled A, B and C. For each transformer, say whether it is a step-up or step-down transformer.

4 Look at Figure B. Suggest why the voltage is reduced before electricity is supplied to homes.

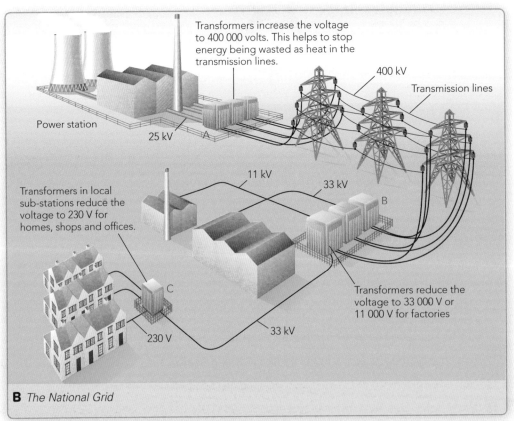

Transformers increase the voltage to 400 000 volts. This helps to stop energy being wasted as heat in the transmission lines.

400 kV

Transmission lines

Power station

25 kV

A

Transformers in local sub-stations reduce the voltage to 230 V for homes, shops and offices.

11 kV

33 kV

B

Transformers reduce the voltage to 33 000 V or 11 000 V for factories

230 V

33 kV

C

B *The National Grid*

The voltage of an alternating current can be changed using a **transformer**. A **step-up transformer** increases the voltage (and decreases the current at the same time). A **step-down transformer** makes the voltage lower and the current higher.

A transformer consists of two coils of insulated wire wound on to an iron core. Electricity is supplied to the **primary coil** and obtained from the **secondary coil** at a different voltage.

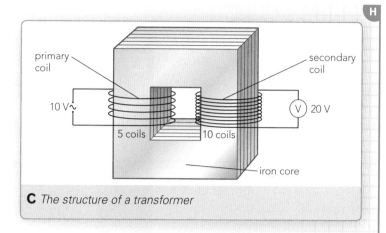

C *The structure of a transformer*

primary coil
secondary coil
10 V∼
(V) 20 V
5 coils
10 coils
iron core

ResultsPlus
Watch Out!

The energy transferred depends on the voltage *and* the current. If the voltage is increased the current is decreased.

D

Danger
Overhead electric power lines

No fishing beyond this point

The potential difference produced by a transformer can be calculated by:

$$\frac{\text{voltage (primary)}}{\text{voltage (secondary)}} = \frac{\text{turns (primary)}}{\text{turns (secondary)}} \quad \text{or} \quad \frac{V_p}{V_s} = \frac{N_p}{N_s}$$

e.g.

For example, a radio runs off the mains supply, but needs only a 23 V supply. The transformer has 100 turns of wire in the primary coil. How many turns are needed in the secondary coil?

$$\frac{V_p}{V_s} = \frac{N_p}{N_s}$$

$$\frac{230V}{23V} = \frac{100}{N_s}$$

Rearranging:

$$\frac{N_s}{100} = \frac{23V}{230V}$$

$$N_s = \frac{23 \times 100}{230} \qquad N_s = 10$$

Electricity supplied to homes is alternating current at 230 V. You can get an electric shock from a 230 V supply, and it might even kill you. The higher voltages used in other lines, such as transmission lines, are even more likely to kill.

You get an electric shock if one part of your body is at a higher voltage than another because this makes an electric current run through you. Birds sitting on electricity transmission lines do not get a shock because both their feet are on the wire at the same voltage. However, if you fly a kite near some pylons, electricity may flow from the high voltage wire through the string and then through you and into the Earth.

?

5 Electricity substations contain transformers. Why shouldn't children play in or near a substation?

6 Look at Figure D. Why is this warning sign necessary?

7 Look at Figure C. If the voltage across the secondary coil on the step-up transformer is 75 V, what is the voltage across the primary coil?

8 Describe the stages in the way electricity is generated and transmitted around the country.

Learning Outcomes

5.11 Recall that a transformer can change the size of an alternating voltage

H **5.12** Use the turns ratio equation for transformers to predict either the missing voltage or the missing number of turns

5.13 Explain why electrical energy is transmitted at high voltages, as it improves the efficiency by reducing heat loss in transmission lines

5.14 Explain where and why step-up and step-down transformers are used in the transmission of electricity in the National Grid

5.15 Describe the hazards associated with electricity transmission

HSW **12** Describe the benefits, drawbacks and risks of using new scientific and technological developments

>>>>>>>>>>>>>>>>>>>>>>>>>> What is a smart meter and how can it help you to save money?

How are electricity bills calculated?

Every home in the UK should have a smart meter by 2020. These meters display how much electricity people are using from the mains supply and how much it is costing them. Electricity bills are worked out from the amount of energy used.

A

The unit of measurement for energy is the **joule** (**J**). The amount of energy that is used by a working appliance each second is known as its **power**. The unit of measurement for power is joule per second or **watt** (**W**) (1 W = 1 J/s).

$$\text{power (watt, W)} = \frac{\text{energy used (joule, J)}}{\text{time (second, s)}}$$

We can write this as symbols:

$$P = \frac{E}{T}$$

The power of an electrical appliance can also be worked out from the voltage and current.

power (watt, W) = current (ampere, A) × voltage (volt, V)

We can write this as symbols:

$$P = I \times V$$

The energy transferred by an appliance depends on its power and how long it is switched on. Since many electrical appliances need a lot of power, appliances are usually given a 'power rating' in **kilowatts** (**kW**) (1 kW = 1000 W) which is shown on the appliance.

1 What does 'power' mean?

2 a A radio uses a 9 V battery and the current is 2 A. What is the power of the radio?
H b How much *energy* will the radio use if it is switched on for one hour?

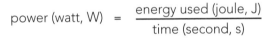

Product Code: 30110
Model :3SLGB
Serial No: 2004.07/294/

Dualit

County Oak Way Crawley West Sussex RH11 7ST
Tel: +44 (0)1293 652500
Email: info@dualit.com
Made in England

Volts :230a.c.
Watts:1700
2076

CE

Electricity companies charge for electrical energy by the **kilowatt-hour** (**kWh**). This is the amount of energy that is transferred by a 1 kW device in 1 hour. Electricity companies usually give their prices in pence per kWh. They call 1 kWh a **Unit** of electricity.

B *This toaster has a power rating of 1.7 kW (1700 W).*

You can work out the cost of electricity using this formula:

cost		power		time		cost of 1 kWh
(p)	=	(kilowatt, kW)	×	(hour, h)	×	(p/kWh)

For example, an electric heater has a power rating of 2 kW. What is the cost of using the fire for 3 hours if one kWh of electricity costs 12 p?
cost in pence = 2 kW × 3 h × 12 p/kWh = 72 p

C The charging unit for this electric car has a power of 4.8 kW. It takes about 4 hours to charge up a car.

1 June 2005

1 September 2005

D Meter readings taken 3 months apart

3 What is a Unit of electricity?

4 Electricity costs 14p per kWh. How much will it cost to have a 3 kW electric heater switched on for 3 hours?

5 The electricity meter in Figure D shows the number of kilowatt-hours used between two dates. If electricity costs 7.5p per kWh, what will the electricity bill be for the 3 months?

H 6 Look at Figure C.
a The charger uses the mains supply. What current does it use?
b If it costs £3.50 to recharge a car, how much does electricity cost per kWh?

7 Compare the meters in Figures A and D. Suggest why standard electricity meters are being replaced by smart meters.

Learning Outcomes

5.2 Define power as the energy transferred per second and measured in watts

5.3 Use the equation: electrical power (watt, W) = current (ampere, A) × potential difference (volt, V)
$P = I \times V$

5.16 Recall that energy from the mains supply is measured in kilowatt-hours

5.17 Use the equation: cost (p) = power (kilowatts, kW) × time (hour, h) × cost of 1 kilowatt-hour, (p/kW h)

5.21 Use the equation: power (watt, W) = energy used (joule, J) / time taken (second, s) $P = \dfrac{E}{T}$

HSW 11 Present information, develop an argument and draw a conclusion using scientific, technical and mathematical language, and ICT tools

P1.29 Power consumption

How much power do different appliances use?

Sports cars can accelerate and travel faster than family cars because their engines are more powerful. However, the driver of the sports car can control the amount of power being used. The further the accelerator is pressed, the more fuel is supplied to the engine each second, and the more power is used.

A *This train can travel at over 580 km/h.*

Figure A shows the fastest electric train in the world – the Japan Rail MLX01, which can travel at over 580 km/h. How is its power controlled?

The power of an appliance is the amount of energy it transfers each second. Power is measured in watts. For electrical appliances, the power can be worked out by measuring the voltage across the appliance and the current flowing through it.

B *These appliances use different amounts of power.*

power (W) = current (A) x voltage (V)

You can find the power of an appliance if you know the current and voltage. An ammeter, placed in series in a circuit, can be used to measure current. The voltage is measured with a voltmeter placed in parallel.

Your task

You are going to plan an investigation that will allow you to find out the relationship between power and voltage in electrical appliances. The hypothesis is that the power of an appliance depends on the voltage supplied to it. Your teacher will provide you with some materials to help you organise this task.

Learning Outcomes

5.4 *Investigate the power consumption of low-voltage electrical items*

Build Better Answers

When completing an investigation like this, one of the skills you will be assessed on is your ability to *evaluate your method*. There are 6 marks available for this skill. Here are two student extracts that focus on this skill. Other skills that you need for the assessment are dealt with in other lessons.

Student extract 1 — A basic response for this skill

This is a good example of clearly pointing out a problem with the method.

Suggest a sensible way to fix the problem.

> It was difficult to carry out the practical sometimes because the ammeter reading kept changing and it was difficult to know what the right reading was. I think that I could make my method better by repeating the readings three times for each different setting and then working out an average reading. This will give me more data to work with and will make it less likely for me to include an anomalous reading by mistake.

Explaining why your change will make your method better is also important.

Student extract 2 — A good response for this skill

You don't just have to put weaknesses – sometimes it is a good idea to say what worked well.

Quality of data means how repeatable your data is, if there are any anomalies and whether it contains enough information for you to prove or disprove your hypothesis.

> The investigation worked well because it gave me a full set of results but the ammeter readings flickered and were difficult to take at times. We have also only tested one appliance so although we can come to a conclusion the hypothesis is only proved for one appliance. The results produced a graph but it was difficult to see the exact shape at some points. In order to improve this method I think that I could take repeat readings and then calculate means. This would improve the quality of the data. I could also take readings at smaller intervals and this would provide more data and allow me to see the full shape of the graph. It would also be a good idea to test different appliances so that we could test the hypothesis fully.

In order to access the higher marks you need to be able to explain how any weakness affected the quality of the data.

It is important to explain how any changes you suggest will relate to the hypothesis.

 ResultsPlus

To access 4 marks
- Describe strengths or weaknesses in your method
- Provide reasons for any anomalies
- Suggest how to improve your method and explain how this will improve the quality of the evidence you could collect

To access 6 marks
You also need to:
- Relate your comments back to the original hypothesis of the investigation

How can we work out the best way to reduce our energy use?

The sale of 100 W light bulbs with filaments was banned in Europe from 2009. They are being replaced by different kinds of low energy light bulbs.

As most of the electricity in the UK is generated using fossil fuels, using electricity adds carbon dioxide to the atmosphere. Most scientists think that this extra carbon dioxide is contributing to climate change. If we use less energy, we reduce the amount of carbon dioxide added to the atmosphere and also save money.

A The Lumiblade lights here use 'OLED' technology to prod light without using much electricity.

Modern appliances do not waste as much energy as older designs. For example, a 10-year-old fridge will use about 85 kWh more electricity per year than a new one. However, not all new fridges use the same amount of electricity, and although one brand may use less energy than another, it may be more expensive to buy.

1 Give two advantages of using low energy appliances.

2 If someone replaced their 10-year-old fridge with a new one, how much money would they save each year if electricity costs 15p per kWh?

3 Why might someone buy a low energy fridge even if it were more expensive?

4 a Give two reasons why a microwave oven would use much less energy than a conventional oven to cook a piece of food.
b What are the benefits and drawbacks of using microwave ovens for cooking instead of using normal ovens?

JACKET POTATOES

You will need:

1 large potato

Cook for:

1½–2 hours in a medium oven

15–20 minutes in 850 W microwave oven

B Cooking times for jacket potatoes

Sometimes you can use a different kind of appliance to do the same job. For example, microwave ovens heat food in a completely different way to normal ovens which means the food cooks much more quickly. Because the microwave oven itself does not get hot, it also transfers less wasted heat energy to the surroundings.

C Different ways of cooking cakes. The one on the left was cooked in a conventional oven and the one on the right was cooked in a microwave oven.

Over half of the energy used in homes is used for heating the rooms and heating water. We could use less energy by making sure our homes are well insulated and by using solar energy to heat water directly. There are different ways this can be done, as shown in Figure D.

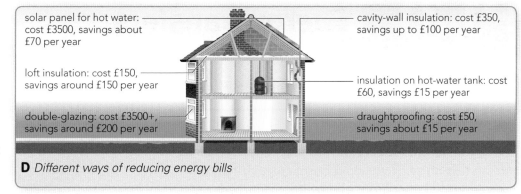

solar panel for hot water: cost £3500, savings about £70 per year

cavity-wall insulation: cost £350, savings up to £100 per year

loft insulation: cost £150, savings around £150 per year

insulation on hot-water tank: cost £60, savings £15 per year

double-glazing: cost £3500+, savings around £200 per year

draughtproofing: cost £50, savings about £15 per year

D *Different ways of reducing energy bills*

As with low energy fridges or light bulbs, all these methods of reducing energy usage cost money. The length of time it takes you to save the amount of money it costs to buy the item is the **payback time**. The most **cost-efficient** method saves the most money compared with its cost. The most cost-efficient method is the one with the shortest payback time.

e.g.

For example, it costs £60 to insulate the hot-water tank, and this will save you £15 a year on energy bills. How long will it take you to save the money on energy bills that it cost you to install the insulation?

$$\text{payback time} = \frac{\text{cost}}{\text{savings per year}}$$

$$= \frac{£60}{£15}$$

$$= 4 \text{ years}$$

?

5 Look at Figure D.
a Which energy saving method is the cheapest?
b Which method will save the most energy? Explain your answer.

6 a Work out the payback times for all the energy saving methods shown in Figure D.
b Which method is the most cost-efficient?

7 Explain why people add extra insulation to their homes and why someone might decide to insulate their loft first, even though it is not the cheapest energy saving method.

Skills spotlight

Scientists working in big companies are looking for new ways of doing things. Look at the lights in Figure A. Suggest why the manufacturers wanted to develop new light bulbs and why people might want to buy them.

ResultsPlus
Watch Out!

To calculate payback time, find the cost of the method used and divide it by the estimated annual saving.

Learning Outcomes

5.18 Demonstrate an understanding of the advantages of the use of low energy appliances

5.19 Use data to compare and contrast the advantages and disadvantages of energy saving devices

5.20 Use data to consider cost-efficiency by calculating payback times

HSW 12 Describe the benefits, drawbacks and risks of using new scientific and technological developments

How do we show energy transfers?

Owls can fly almost silently, and aircraft engineers are trying to mimic owl wings to see if they can make aircraft less noisy. They could design aircraft that will convert less of their input energy into sound energy or design aircraft parts that will convert any sound energy produced into other forms of energy.

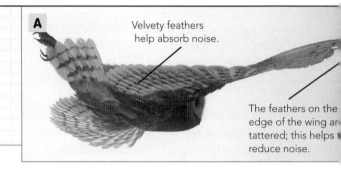

A Velvety feathers help absorb noise.

The feathers on the edge of the wing are tattered; this helps to reduce noise.

> **1** What do we mean by 'energy is conserved'?
>
> **2** What are the input energy and output energy forms for a torch?

Energy transfers

There are nine forms of energy: **thermal** (also called heat energy); **light**; **electrical**; **kinetic** (also called movement energy); **sound**; **chemical potential** (such as the energy stored in batteries and fuels); **nuclear potential** (the energy stored in the nuclei of atoms); **elastic potential** (the energy stored by things that have been stretched or squashed and can spring back) and **gravitational potential** (the energy stored in things that can fall).

Energy can be moved from one place to another; this is called **energy transfer**. When this happens, energy is often transformed from one form into another.

Conservation of energy

In physics, a **system** is the word used for something in which we are studying changes. Systems can be very complex (e.g. a planet) or quite simple (e.g. a piece of metal).

If you add up all the energy that has been transferred by a system (the output energy) and compare it with the energy put into the system (the input energy), the amounts are the same. So we can say:

output energy = input energy

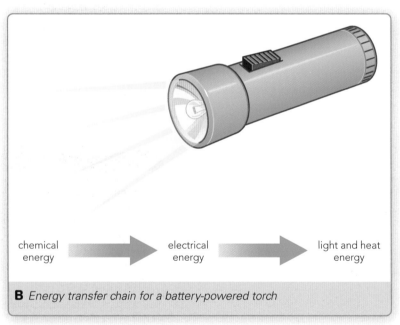

chemical energy → electrical energy → light and heat energy

B Energy transfer chain for a battery-powered torch

This means that energy cannot be created or destroyed. It can only be transferred from one place to another or transformed from one form to another. This is called the **law of conservation of energy**.

> **3** Draw an energy transfer chain for:
> **a** a radio
> **b** a computer
> **c** a bouncing ball.

Although energy is always conserved, it is not always transferred into forms that can be used. Think of a bouncy ball. When the ball bounces, its temperature rises slightly, and its kinetic energy after the bounce is a little less than it was before. This means the ball will bounce to a lower height than it was dropped from.

Energy conversion diagrams

An energy conversion diagram (sometimes called a Sankey diagram) shows the amount of energy converted or transferred. The width of the arrows represents the amount of energy in joules.

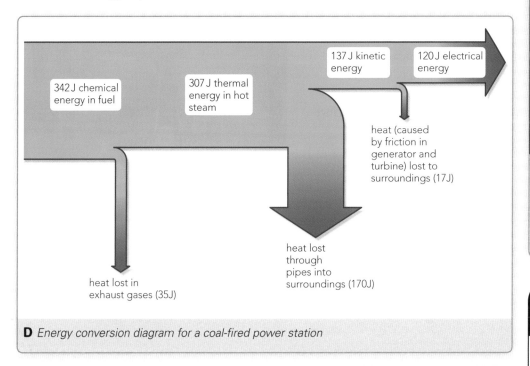

137 J kinetic energy

120 J electrical energy

342 J chemical energy in fuel

307 J thermal energy in hot steam

heat (caused by friction in generator and turbine) lost to surroundings (17J)

heat lost through pipes into surroundings (170J)

heat lost in exhaust gases (35J)

D Energy conversion diagram for a coal-fired power station

4 An electric drill converts 1000 J of electrical energy into 580 J of kinetic energy and 20 J of sound energy. The remaining energy passes to the surroundings as thermal energy. Calculate how much thermal energy is produced and draw an energy conversion diagram for the drill.

5 a Choose an electronic device and write an energy transfer chain for it.
b Make up values for energy input and energy output for this device and draw an energy conversion diagram to show them.

6 Draw a diagram to illustrate the law of conservation of energy.

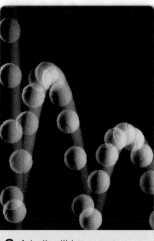

C A ball will bounce to a lower height than it is dropped from because kinetic energy is transferred to thermal energy.

Results Plus
Watch Out!

In an energy conversion diagram the output energy arrows must add up to the amount of input energy.

Skills spotlight

Scientists often find it useful to present technical information in mathematical ways. Why is an energy conversion diagram more useful than an energy transfer chain?

Learning Outcomes

6.1 Demonstrate an understanding that energy is conserved

6.2 Describe energy transfer chains involving the following forms of energy: thermal (heat), light, electrical, sound, kinetic (movement), chemical, nuclear and potential (elastic and gravitational)

6.3 Demonstrate an understanding of how diagrams can be used to represent energy transfers

HSW 11 Present information, develop an argument and draw a conclusion, using scientific, technical and mathematical language, and ICT tools

P1.32 Efficiency

What is energy efficiency?

When a new car is being developed, prototypes are designed and used to test certain ideas. The VW L1 prototype was part of the company's attempt to build a highly fuel-efficient but useable car. It was capable of using just one litre of diesel per 100 kilometres travelled and technologies developed for the car have now been used in other models.

A *The Volkswagen L1 prototype*

1 a Describe the energy transfers in a car.
b Which forms of energy in **a** are useful?
c Which forms of enegy in **a** are wasted?

Wasted energy

When a light bulb is switched on, most of the electrical energy supplied to it is converted into thermal energy. This energy spreads out to the surroundings and cannot be used for other useful energy transfers – it is wasted.

Efficiency

2 Why do you think old-style 'filament' light bulbs were banned?

How good a device is at converting energy into useful forms is known as its efficiency. Low energy light bulbs transform more of the input electrical energy into light energy than older-style bulbs. They are more efficient.

Maths skills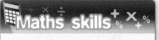

The energy efficiency grades in Figure D are based on **percentiles**.

A sample of appliances is taken and ranked by their energy efficiency. The sample is then split into 100 equally sized groups. The group with the highest energy efficiencies is the 100th percentile and the group with the lowest energy efficiencies is the 1st percentile.

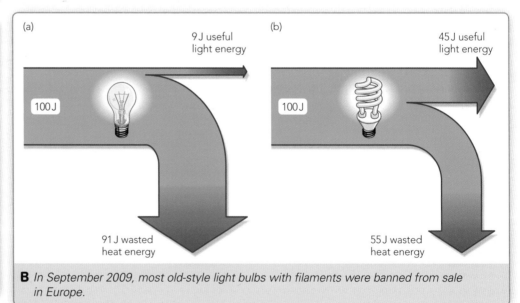

(a) 9 J useful light energy — 100 J — 91 J wasted heat energy

(b) 45 J useful light energy — 100 J — 55 J wasted heat energy

B *In September 2009, most old-style light bulbs with filaments were banned from sale in Europe.*

We can use an equation to calculate the percentage efficiency of a device:

$$\text{efficiency} = \frac{\text{useful energy transferred by the device}}{\text{total energy supplied to the device}} \times 100\%$$

For example, for 200 J of input energy, a jet pack produces 80 J of kinetic energy, 10 J of sound and 110 J of thermal energy. Calculate its efficiency.

useful energy transferred = 80 J

total energy supplied = 200 J

$$efficiency = \frac{80\,J}{200\,J} \times 100\% = 40\%$$

ResultsPlus
Watch Out!

Make sure you use the efficiency equation with the correct energy amounts in the right places. If your answer is more than 100% then you have made a mistake!

Household goods are given an energy rating. Rather than give these as percentages, they are given as a colour coded grade from A to G, with A being the most efficient.

C A more sufficient jet pack will allow it to carry you further.

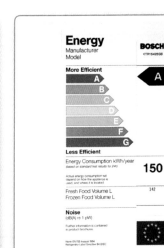

D Energy efficiency ratings allow you to compare the efficiency of electrical devices.

(?)

3 What are the efficiencies of the two light bulbs shown in Figure B?

4 What would you need to know in order to calculate the efficiency of the L1 prototype car and another car, in order to compare them?

Skills spotlight

When scientists want to answer a question, they often plan an experiment that will provide data. Write a plan for an experiment to determine the efficiency of an electric kettle. Show how the data you collect will help you calculate the kettle's efficiency.

(?)

5 A radio is supplied with 50 J of electrical energy and converts this into 5 J of sound energy with the rest wasted as heat.
a How many joules of heat does the radio transfer to the surroundings?
b What is the efficiency of the radio?

6 Explain what the letter ratings on a new fridge mean and discuss whether you think these ratings are useful. 🖊

Learning Outcomes

6.4 Apply the idea that efficiency is the proportion of energy transferred to useful forms to everyday situations

6.5 Use the efficiency equation:

$$efficiency = \frac{useful\ energy\ transferred\ by\ the\ device}{total\ energy\ supplied\ to\ the\ device} \times 100\%$$

HSW **5** Plan to test a scientific idea, answer a scientific question, or solve a scientific problem by selecting appropriate data to test a hypothesis

>>>>>>>>>>>>>>>>>>>>>>>>

Why might a pretty teapot serve cold tea?

P1.33 Heat radiation

Which colours radiate heat the quickest?

Scientists are researching different surfaces for car bodies to make them more efficient. (An efficient car will go further using less fuel.) They are even trying to find out if different colours make a difference.

The car's radiator also needs to be as efficient as possible at removing heat from the engine. If an engine gets too hot, it can seize up. One way of increasing the efficiency of the radiator is to increase its surface area. But does its colour also make a difference?

A This car is covered in dimples, like shark skin and golf balls, and the makers claim it makes the car more efficient.

On a sunny day, a car will absorb (take in) infrared radiation from the Sun. When the Sun has set, the car will radiate the thermal energy (as infrared radiation).

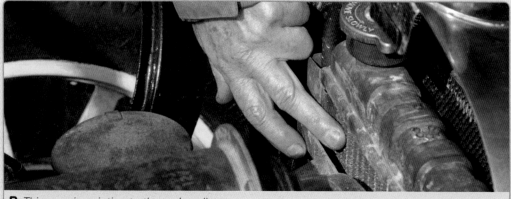

B This man is pointing to the car's radiator.

This can make the car feel warm. Black cars heat up more than other colours of car because they absorb more radiation from the Sun.

Black is the best colour at absorbing thermal energy, but is it also the best colour at radiating it? Would another colour radiate heat better?

Your task

You are going to plan an investigation to find out what colour radiates heat the fastest. You are going to test the hypothesis that black is the best colour for a car radiator. Your teacher will provide you with some materials to help you organise this task.

Learning Outcomes

6.7 Investigate how the nature of a surface affects the amount of thermal energy radiated or absorbed

Build Better Answers

When completing an investigation like this, one of the skills you will be assessed on is your ability to *identify and control variables*. There are 6 marks available for this skill. Here are two student extracts that focus on this skill. Other skills that you need for the assessment are dealt with in other lessons.

Student extract 1 — A basic response for this skill

> In this practical I need to control the amount of water I use each time and whether the containers have a lid or not and the amount of light which falls on the container. I will use a measuring cylinder to measure the same volume of water each time. I will make sure that each container has a lid of some sort.

Make sure that you are specific in what you say. In this case it might be better to make the lids from the same material.

Light is not relevant to this practical.

Describe how you are going to control each variable you mention.

Student extract 2 — A good response for this skill

If you give the full range of relevant variables then you can access the highest marks.

> There are several different variables I will need to control in this practical. I will need to control the amount of water I use each time, the starting temperature of the water, the surface of the containers and whether they have lids or not. I will do this by using a measuring cylinder to measure 100cm³ of water each time. A measuring cylinder will mean that I can be sure that I am measuring the water properly each time. I will control the starting temperature each time by using a thermometer and not starting the stopwatch until it shows that the water is at 85°C. By using a thermometer I can be sure that the temperature is the same each time. I will control the surface of the containers by using a selection of differently coloured plastic cups. This will mean that each cup is made of the same material and has the same surface but the only difference will be the colour. This will remove the possibility that the surfaces are affecting the radiation of energy. I will use the same style of plastic lid each time. This will ensure that heat loss through convection is limited as much as possible.

To work at the highest level you need to include a description of how to control the variable and an explanation of why you are going to do it that way.

ResultsPlus

To access 4 marks
- Identify some relevant variables to control
- Describe how to appropriately control these variables

To access 6 marks
- Identify a range of relevant variables to control
- Explain how to control these variables

How do you keep a swimming pool at constant temperature?

A

At the Beijing Olympics in 2008, the new swimming pool design helped athletes break 25 world records. The pool was maintained at 27°C, the best temperature for speed swimming.

1 A plate of food on a hotplate absorbs power at a rate of 32 W. What must happen for it to stay at 50°C?

Constant temperatures

The amount of energy transferred in a certain time is the power. It is measured in watts (W) (1 W = 1 J/s). For a system to stay at a constant temperature it must absorb the same amount of power as it radiates. If a swimming pool at 27°C radiates 1200 W, it needs 1200 W transferred to it by the heating system in order to stay at 27°C.

Skills spotlight

Difficult ideas can sometimes be made clearer using a model. A laboratory water bath could be used as a model for how the Earth keeps a constant temperature. Evaluate this model and say whether you think it is a useful one or not.

Earth's energy balance

The Sun radiates energy. The Earth's surface absorbs about half of the radiation that reaches the Earth and re-radiates this energy (as infrared radiation, which can heat up the atmosphere). For the temperature of the Earth to stay the same, it must radiate energy into space at the same average rate as it absorbs it.

In Figure B the power absorbed by the Earth and its atmosphere is the same as the power radiated. So the system stays at a constant temperature.

So-called 'greenhouse gases', such as carbon dioxide in our atmosphere naturally absorb some energy and keep the Earth at a higher temperature than if there were no atmosphere. However, many scientists believe that human activities have upset this balance and that the Earth is getting warmer because of an increase in greenhouse gases.

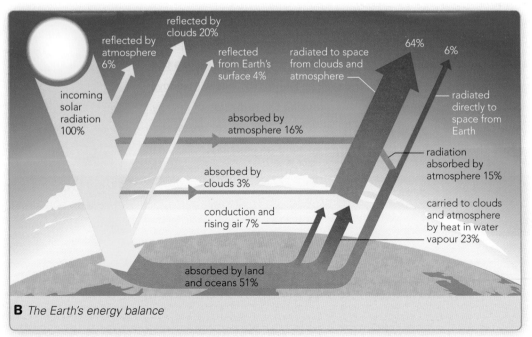

reflected by clouds 20%
reflected by atmosphere 6%
reflected from Earth's surface 4%
radiated to space from clouds and atmosphere
64%
6%
incoming solar radiation 100%
absorbed by atmosphere 16%
radiated directly to space from Earth
radiation absorbed by atmosphere 15%
absorbed by clouds 3%
conduction and rising air 7%
carried to clouds and atmosphere by heat in water vapour 23%
absorbed by land and oceans 51%

B The Earth's energy balance

There is only a certain amount of heat that additional greenhouse gases absorb. So if greenhouse gas emissions are stopped then the increase in temperature will also stop, and the Earth and its atmosphere will have reached a new (but higher) constant temperature.

If greenhouse gases are removed from the atmosphere, the atmosphere will be able to hold less energy and its temperature will decrease.

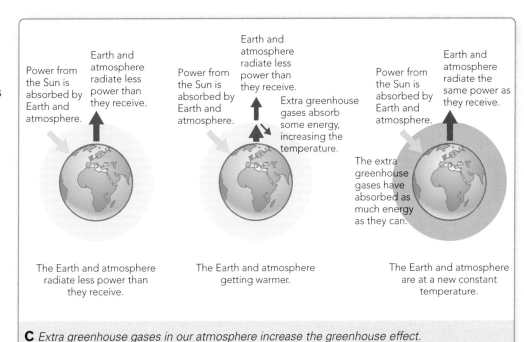

Power from the Sun is absorbed by Earth and atmosphere.

Earth and atmosphere radiate less power than they receive.

The Earth and atmosphere radiate less power than they receive.

Power from the Sun is absorbed by Earth and atmosphere.

Earth and atmosphere radiate less power than they receive.

Extra greenhouse gases absorb some energy, increasing the temperature.

The Earth and atmosphere getting warmer.

Power from the Sun is absorbed by Earth and atmosphere.

Earth and atmosphere radiate the same power as they receive.

The extra greenhouse gases have absorbed as much energy as they can.

The Earth and atmosphere are at a new constant temperature.

C *Extra greenhouse gases in our atmosphere increase the greenhouse effect.*

Various scientists have suggested ideas to change the power absorbed and radiated by the Earth and so stop the Earth's temperature rising. We could place huge white screens in space, about 2000 km along each side. These would reflect sunlight and shade the Earth. Another idea is to float millions of white ping pong balls on ocean surfaces to reflect more sunlight.

Results Plus
Watch Out!

Joules measure energy and watts measure power.

2 On average, each square metre of the Earth's surface could receive 343 W of solar power. Some of this energy doesn't reach the surface – it is absorbed or reflected by the atmosphere.
a How much power is absorbed by each square metre of Earth on average?
b How much power, re-radiated from each square metre, goes directly into space?
c What would happen if less than this amount went into space?

3 If the Earth's average temperature rises to a new steady level, what can you say about the power absorbed and radiated by the Earth and atmosphere?

4 Explain what effect giant mirrors pointing at the Sun would have on the temperature of the Earth.

5 Ping pong balls can be made in different colours. What colour would be best for reducing the Earth's temperature rise? Explain your answer.

6 Explain with a simple example how any system can stay at a constant temperature.

Learning Outcomes

6.6 Demonstrate an understanding that for a system at a constant temperature it needs to radiate the same average power that it absorbs

HSW 3 Describe how phenomena are explained using scientific models

Uses of the electromagnetic spectrum

1. All regions of the electromagnetic spectrum have some uses.

   (a) (i) Microwaves can be used for

   A communication
   B detection of cancer
   C security scanners
   D TV remote control   (1)

   (ii) Gamma rays can be used for

   A communication
   B detection of cancer
   C security scanners
   D TV remote control   (1)

   (b) Explain how ultraviolet lamps are used in detecting forged banknotes.   (2)

   (c) SODIS (solar water disinfection) is a system that allows people in countries with poor water supplies to disinfect their drinking water.

   The photograph shows some villagers placing bottles out in the sunshine. After six hours in the sunshine, the water is both heated and disinfected.

   (i) What does disinfected mean?   (1)

   (ii) Name the part of the electromagnetic spectrum which will:

       (a) disinfect the water

       (b) heat the water   (2)

   (iii) The water used should be clear not cloudy. How would cloudy water prevent the water from being disinfected?   (2)

Did the Earth really shake?

2. Rachelle places a magnet in a coil of wire attached to a data logger.

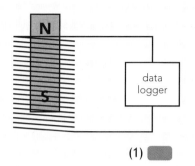

(a) (i) When the magnet and coil are not moving the data logger will:

   A read zero
   B decrease
   C increase
   D change direction

   (1) ▢

   (ii) When Rachelle pulls the magnet upwards the data logger gives a positive reading.
   To give a bigger reading on the data logger the speed of the magnet must:

   A stop moving
   B move slower
   C move faster
   D change direction

   (1) ▢

(b) A geophone can measure the ground movement caused by an earthquake.

The magnet is fixed inside a frame. The coil hangs from the frame by springs. The frame is fixed to the Earth. If the Earth moves, the magnet moves inside the coil.

The current reading indicates the speed of movement of the ground. The graph shows how the speed of the ground changes as earthquake waves pass.

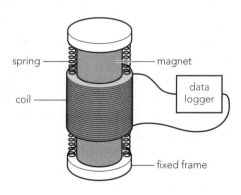

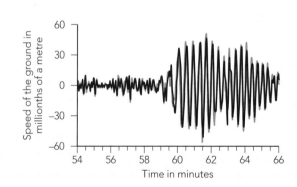

(i) Estimate the frequency of the earthquake wave. (2) ▢

(ii) Describe what happens to the current in the geophone after 59 minutes. (2) ▢

(c) The current in the geophone is directly proportional to the speed of the ground.
Sketch a graph to show this relationship between current and speed. (2) ▢

Energy and the Earth

3. The diagram below shows what happens to energy arriving at the Earth from the Sun.

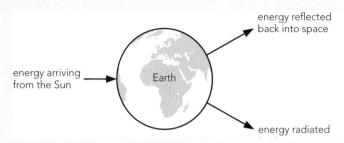

(a) (i)   Visible light arrives at the Earth's surface from the Sun. Name two other types of
radiation which reach the Earth's surface.   (2)

(ii) Imagine that the overall temperature of the Earth remains constant. Compare the
amounts of energy the Earth **receives and returns** into space in the same time.   (1)

(iii) On average, for each 100 joules of energy the Earth receives, it radiates 70 joules back
to space. How much energy is **reflected** back into space?   (1)

(b) The diagram below gives more detail about the energy flows.

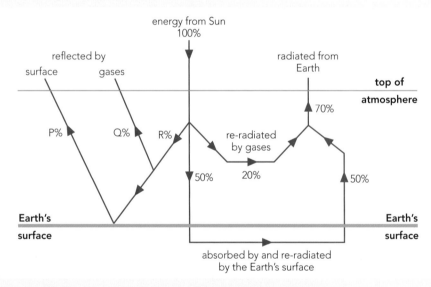

(i) Make an equation to show the relationship between P, Q and R.   (1)

(ii) The value of Q is five times the value of P. Find the value of R and then calculate the
numerical value of P.   (3)

(c) Many people say that the overall temperature of the Earth is changing. Explain, using
information from the diagrams above, how cutting down trees is increasing the overall
temperature of the Earth.   (2)

Telescopes

4. Telescopes have been widely used for over 400 years. They are an important tool that help astronomers discover more about the Universe.

   (a) (i) Telescopes with two lenses are best described as

         A compound
         B radio
         C reflecting
         D refracting   (1) ▢

   (ii) In the 1930s, scientists made the first telescope that used radio waves instead of light waves. Describe one way in which a radio telescope is different from a telescope that uses light.   (1) ▢

   (b) John uses a lens to focus the light from the Sun onto a piece of paper. The paper soon starts to burn. What is the distance between the paper and the lens called?   (1) ▢

   (c) State **two** ways in which photographing stars is better than simply looking through a telescope.   (2) ▢

   (d) The picture shows an astronomical telescope.

lens A   lens B

   Toni makes a model of this telescope using a thin lens and a fat lens. Explain how the two lenses can be arranged to form a virtual image of a distant tree when she looks through the eyepiece. Your answer should explain why the image is virtual.   (6) ▢

Earthquakes

5. (a) (i) Which row of the table is correct for seismic S-waves? (1)

	Can they be reflected?	Can they be refracted?
A	no	no
B	yes	no
C	no	yes
D	yes	yes

(ii) The Earth's outermost layer is composed of plates moving near or against each other. How does this movement of plates cause earthquakes? (2)

(b) Earthquakes can produce both longitudinal and transverse waves. Draw diagrams to show the difference between these types of waves. (2)

(c) Earthquakes happen when the tectonic plates move. John investigates sliding as a model for earthquakes. His apparatus is shown in the diagram.

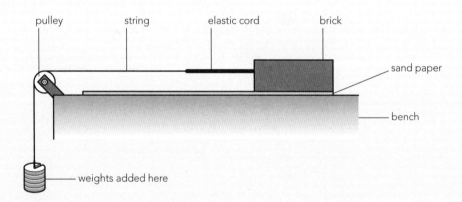

Describe how John could use this aparatus to investigate sliding movements of the brick. What variables will he need to measure, and what variables must be controlled? How does this investigation relate to what happens in real earthquakes? (6)

Models of the Solar System

1. For thousands of years, people have tried to discover what our Solar System is like. There are two main models: the geocentric model and the heliocentric model.

   (a) (i) Which of these statements apply to the heliocentric model of the Solar System?

        A   The planets orbit the Earth.
        B   The Sun orbits the Earth.
        C   The planets orbit the Sun.
        D   The comets orbit the planets.   (1) ▭

     (ii) Which of these is true for both the geocentric and the heliocentric models of the Solar System?

        A   The Moon orbits the Earth.
        B   The Earth orbits the Moon.
        C   The Sun orbits the Earth.
        D   The Earth orbits the Sun.   (1) ▭

   (b) In the 17th century, Galileo used a telescope to observe Jupiter.

     (i) Describe the observations he made.   (2) ▭

    (ii) Suggest why these observations were important.   (2) ▭

   (c) The table and diagram show some data on distances in the Solar System.

		distance
Sun to Earth	B	1.5×10^{11} m
Sun to Jupiter	A	7.8×10^{11} m
Earth to Jupiter		

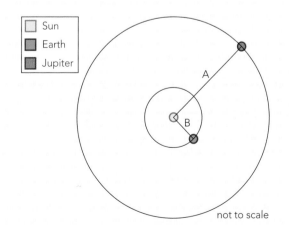

not to scale

Calculate the time taken for light to travel from Jupiter to the Earth when Jupiter and the Earth are at their closest approach. The speed of light is 3×10^8 m/s.   (4) ▭

Infrared radiation

2. Infrared radiation is an important region of the electromagnetic spectrum.

    (a) (i) All electromagnetic waves have the same
        A frequency
        B speed
        C wavelength
        D amplitude   (1)

    (ii) Name **one** region of the electromagnetic spectrum with a frequency less than infrared.   (1)

    (iii) Name a region of the electromagnetic spectrum which can cause cell mutation.   (1)

    (b) An important use of infrared lasers is for cleaning old photographs. Scientists used two identical lasers for different lengths of time. They measured the thickness of dirt removed with one pulse of each laser. They repeated the experiment with different energy levels. The graph shows their results.

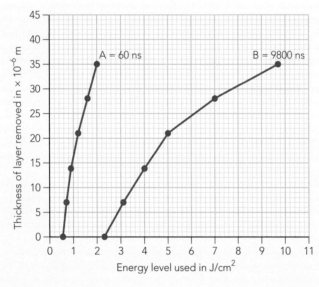

A = 60 nanosecond (ns) pulse
B = 9800 nanosecond (ns) pulse

    (i) How much energy is needed to remove 15×10^{-6} m of dirt with a laser of pulse length 60 ns?   (1)

    (ii) The scientists defined the cleaning ratio as

$$\text{Cleaning ratio} = \frac{\text{thickness removed}}{\text{energy level}}$$

    Calculate the cleaning ratio for laser B at an energy level of 4.8 J/cm².   (2)

    (iii) Explain whether laser A or laser B has the greatest cleaning ratio at low energy levels.   (2)

Waves in the Earth

3. (a) Two laboratories, **P** and **Q**, are used to detect earthquakes. A scientist at each laboratory works out the distance to an earthquake. Then they draw arcs of circles to locate the earthquake.

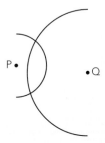

(i) Why are scientists not certain of the position of the earthquake? (1) ▪

(ii) **P** and **Q** are 400 km apart. The earthquake is 500 km from **P** and 200 km from **Q**.

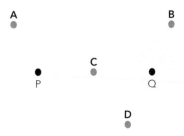

At which position, **A**, **B**, **C** or **D** could the earthquake have happened? (1) ▪

(b) Rasheeq studied earthquakes. He learnt that the density of rock is proportional to the wave speed. The graph shows the speed of two seismic waves, **M** and **N**, at different depths below the Earth's surface.

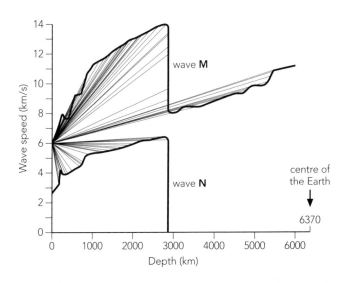

(i) For wave **M**, the average speed at depths less than 2900 km is 11 km/s. How long would it take to travel from the surface to a depth of 2900 km and back again? (3)

(ii) Why do both graphs have a vertical part at 2900 km? (2)

(iii) Why does the speed of wave **M** increase at depths greater than 2900 km but the speed of wave **N** does not? (2)

Distributing electricity

4. The photo shows how electricity is distributed around the country.

(a) (i) The generator at the power station produces current which is

    A alternating
    B direct
    C random
    D unchanging (1)

(ii) To transmit 20 MW of power between two towns, a voltage of 40 000 V is sent across the wires. Calculate the current in the wires. (3)

    $P = I \times V$

(b) High potential differences (voltages) are used for transmitting electrical energy from power station to consumer.

Discuss the advantages and disadvantages of using high voltages instead of domestic and whether the benefits outweigh the drawbacks. (6)

Observing the Universe

5. Our Universe contains many different stars and galaxies.

(a) What is the difference between a galaxy and a star? (1) ▢

(b) Some telescopes are placed outside the atmosphere so that

    A   they are not affected by air resistance
    B   they pass overhead twice in every hour
    C   they can observe stars even when it is cloudy
    D   their solar cells create electricity more efficiently (1) ▢

(c) Describe **one** method, apart from the use of telescopes, by which scientists have gathered data about the planet Mars. (2) ▢

(d) Describe **two** ways in which modern telescopes have improved our understanding of the Universe. (2) ▢

(e) The Big Bang and the Steady State are two common theories to explain the expansion of the Universe. Compare these two theories and the evidence which supports them. (6) ▢

Here are three student answers to the following question. Read the answers together with the examiner comments around and after them.

Electromagnetic spectrum **Grade** **G–D**

The chart shows the electromagnetic spectrum.

radio waves	microwaves	infrared	visible light	ultraviolet	X-rays	gamma rays

The diagram shows three identical thermometers placed in a spectrum of visible light. The scientist Herschel did a similar experiment and discovered the infrared part of the electromagnetic spectrum.

There is a pattern in the temperature readings shown in the diagram.

Suggest how Herschel used the pattern to make his discovery. (6)

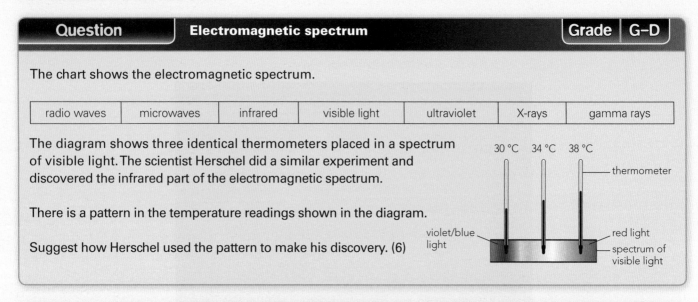

Student answer 1   Extract typical of a level ① answer

This is a very simple attempt at a pattern statement. Use the scientific words from the diagram or chart to improve this.

> Herschel repeated the same experiment as this one. The pattern is low on the left and high on the right. It shows going up on the right. He discovered infrared rays which can't be seen.

Marks are awarded for saying that infrared rays can't be seen, but not for saying that Herschel discovered them – this information is in the question.

Examiner summary
There is not very much new information in this answer, which is needed in order to get full marks. The answer could be improved by interpreting the diagram and charting and linking the temperatures to the colours shown. This would allow the student to identify the pattern shown in the diagram.

Student answer 2   Extract typical of a level ② answer

This simple statement does not show a trend or pattern but the second sentence does clarify it better.

> The pattern is when it is 30°C it is blue and when it is 38°C it is red. It shows that when we move towards the red it is hot. Herschel did this experiment. He used the same equipment as we did in the diagram and he put his thermometer next to the red and it got hot as well. This was his result. He knew he had found a new type of radiation - infrared.

This is a good attempt to describe the pattern, but it could be more clear by writing *the pattern* shows that when *the colour* is red, the thermometer is hot.

It is not clear where 'next to the red' actually is.

Examiner summary
This answer shows an attempt to link all the information given into a pattern or trend but the first sentence really only describes what happens at the extremes. The second sentence is almost enough to explain the trend. The answer could be improved by writing more clearly and in enough detail about the important extra bit that Herschel did in his experiment because 'next to the red' could mean in the orange part of the spectrum or in the infrared. This answer could also say that the fourth thermometer showed a higher temperature.

Some extra detail not mentioned in the diagram is added here. This is good practice as long as the detail is relevant.

This pattern sentence is correct but does not use scientific vocabulary.

Herschel did the same experiment. He used thermometers with blackened bulbs. He made a spectrum with light from a prism and he put the thermometers into the spectrum, one in the blue light and one in the green light and one in the red light. He noticed that the further along into the red the thermometer was the higher the temperature it showed. He wanted to find out if there was anything else next to the spectrum. He put another thermometer further along where he could not see any light to see if the pattern continued. And he found that there was a lot of heat on the thermometer where he could not see any light. He thought this must be because of invisible radiation and he called it infrared. The pattern was the longer the wavelength the more the temperature rises and there was new radiation outside the part of the spectrum we can see.

The detail here locates the fourth thermometer appropriately.

This pattern sentence is clear and uses scientific vocabulary correctly.

Examiner summary

Extra scientific detail has been added here – using blackened thermometers – which shows good familiarity with this experiment. The account is in a logical order. The last sentence seems to have been added to show understanding of what the pattern should be, in case the previous pattern statement was not clear enough. This is good. Although the answer does not explicitly state that the temperature rise is greater in the infrared region compared to the red region, saying a lot of heat is almost enough. The student has clearly linked the lack of visible light to a temperature rise. The last sentence shows understanding of the importance of this discovery.

 ResultsPlus

Move from level ① to level ②

- Use all the information that you are given, especially from diagrams and charts.
- Write pattern statements in the form 'the bigger the..... the greater the.....' e.g. the nearer to the red the hotter the thermometers.
- Make sure that you answer with enough detail to show you know and understand the question.
- Make your answers long enough; this is worth up to six marks, so you should be writing for about six/seven minutes.

Move from level ② to level ③

- Read through your work and make sure that you have no ambiguities in your writing. If necessary add an extra statement at the end.
- You can always include a sketch if you think that will help e.g. to show where the other thermometer is placed.
- Add in extra detail which shows that you know the work.
- It is good practice to write in sentences with correct scientific vocabulary.
- Always write a conclusion or reach a decision or write a statement which shows the importance of the topic you have been writing about.

Here are three student answers to the following question. Read the answers together with the examiner comments around and after them.

Question — Life on Earth
Grade — G–D

The diagram shows how Earth and Jupiter orbit the Sun. Europa, one of the moons of Jupiter, is also shown. The radii of the orbits a, b and c are given in the key.

Some scientists think that there could be simple life on Europa. We have already sent successful manned expeditions to our Moon and there have been plans to travel to Mars. There is now a proposal to send a manned expedition to Europa. Using evidence and opinions, suggest why this is unlikely to happen in the near future. (6)

Earth's orbit

b

a

Sun

c

Jupiter

Europa

not to scale

a = 150 million km

b = 779 million km

c = 1 million km

Student answer 1 — Extract typical of a level ① answer

This is incorrect. Europa is close enough for a spacecraft to reach within a lifetime.

> I think they won't go to Europa in the near future. You would die before you got there. Jupiter is too far away. It would be hard to get there.
>
> I don't think it will happen.

It is important to back up an opinion by giving reasons for it.

Examiner summary
This answer uses informal, everyday English, which doesn't give enough detail to the science ideas. To get more marks it needs to include more facts and information. Opinions must be backed up with scientific facts.

Student answer 2 — Extract typical of a level ② answer

It would be better to be more specific – it would take years to get there.

It would help to include more detail on what life support would be needed on the spaceship.

> Some people think we will not go to Europa soon because it would cost too much and we don't have the right technology. Europa is near Jupiter and it is a long way away. It would take a long time to get there. You would need a big spaceship that you could live on. You would have to take oxygen because there is no air in space.

There are several good ideas here, but the student could go further to talk about the types of technology they mean.

Examiner summary
This is a reasonable answer that mentions several scientific ideas. Most of the statements are linked into a logical sequence showing that you have done some planning, which is helpful. Including more detail in the answer would earn more marks.

Europa is near the planet Jupiter. To get there we would have to use a space ship because we have to travel through a vacuum. It would probably take years because it is further than going to Mars which takes about 10 months. The spaceship would have to have life-support systems like food oxygen and water and somewhere for the astronauts to exercise so they stayed fit. It would cost a lot of money to make and launch such a space ship and I don't think many people would want to go because they would be cut off from their family for so long. It would take hours if not days to get a message from the spaceship back to us, and what if something went wrong? In outer space there would be no chance of a rescue. It might be exciting and interesting to go and we might learn a lot of new things. But there could be lots of better ways of spending our money like research for cancer.

It is good to use link words such as so or because to help expand the detail of an answer.

This tries to think about the other side of the argument, which shows balance.

Each statement is clearly supported with a sensible reason.

It is a good structure because it leaves the opinion to the end so the important facts are covered if time runs out.

Examiner summary

This answer considers the question carefully and gives reasons to back up the statements of fact. The student has tried to cover all the main ideas in enough detail without writing for too long about any one of them. The answer is balanced because it addresses both sides of the situation. Further marks could be gained by using more formal English, but the style is good.

 Results**Plus**

Move from level ❶ to level ❷

- Plan your answer before you write.
- Read your plan through and see if you have your answer in the correct order.
- When you give a personal opinion, e.g. 'I think that...', make sure that you have a 'because...'

Move from level ❷ to level ❸

- Structure your answer as carefully as you can.
- Leave opinion to the end of the answer in case you run out of time.
- Try to write a reason or give more detail for all your statements of facts. This can be done by using linking words such as 'so', 'because' or 'therefore'.
- See if there are any suggestions from the other point of view.

Here are three student answers to the following question. Read the answers together with the examiner comments around and after them.

Question	Earth's structure	Grade	D–A*

The classification chart shows the parts of the Earth through which longitudinal (P) and transverse (S) waves will travel.

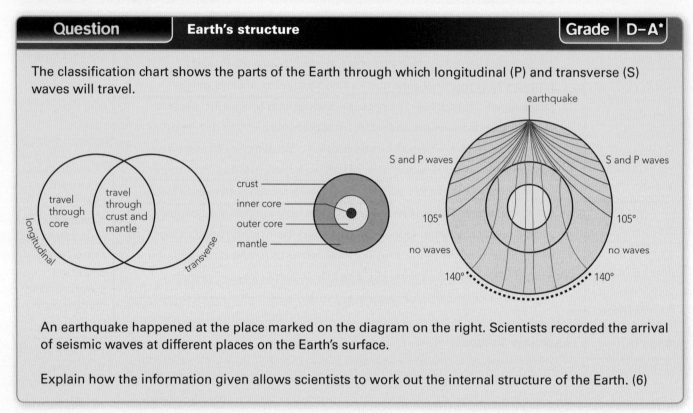

An earthquake happened at the place marked on the diagram on the right. Scientists recorded the arrival of seismic waves at different places on the Earth's surface.

Explain how the information given allows scientists to work out the internal structure of the Earth. (6)

Student answer 1	Extract typical of a level ① answer

This is correct but doesn't answer the question, so it gets no marks.

This correctly identifies which waves travel in which materials.

Avoid using both the name and letter. Here the student has mixed them up.

In an earthquake different parts of the Earth slide against each other. P and S waves move in solids. Both arrive near the earthquake. Only P can go through liquids. The waves bend because the waves go faster and slower. To find how deep the core is find the wave which just misses it. The shadow zone stops any waves passing to the surface. Longitudinal S waves pass right through the liquid core to the south pole but the others don't.

This is unclear. Instead, say the wave travels faster as it moves deeper.

The shadow zone does not stop waves.

Examiner summary

This answer correctly tries to include a comment about each of the points needed. The student also attempts to describe how to find depth of the core layer. However, the answer doesn't use the angles given to help the description. The student needs to try to link the type of waves passing and the nature of the layers. It is important to use technical terms like 'shadow zone', but they must be used correctly.

It would be better to use the specific angles given.

Useful for further evidence, but it would have been better to talk about the different layers.

Starting from the epicentre of the earthquake both longitudinal and transverse waves curve to reach the surface. The longitudinal and transverse waves pass through the centre. At about 60% of the way round no waves get to the surface. This is the shaded zone A wave which just skims the core passes to 105°. A just lower wave refracts into the core and reaches the surface at 140°. No waves reach the surface between these two waves. These two angles allow scientists to work out the size of the core. They can also measure the time to receive waves reflected from the core back to the epicentre.

This is a good use of a technical term.

The correct term is shadow zone.

This could be improved by specifying the type of wave(s).

More detail would be useful.

Examiner summary

The ideas here are explained quite clearly. Use of the term 'shaded' instead of 'shadow' could be excused if it was explained that the zone is a region, not a single place, or what causes it. Using an estimated percentage instead of the angles provided is not the best approach and may cause confusion. To turn this into a level 3 answer, details about refraction, size, and the nature of the layers need to be included.

Not a technical term but descriptive and explained in the next sentence.

This gives a partial description only of how to find the size of core.

Longitudinal and transverse waves both pass through the mantle. This is molten rock but stretchy. This provides the elasticity needed for transverse waves to pass. The core is molten iron and is quite runny. Transverse waves cannot pass through a liquid. As seismic waves travel, they are refracted because they move faster as they go deeper. If a wave curves through the mantle just above the core, it meets the surface 105° from the earthquake. A wave just below this meets the core, slows down and so bends towards the normal. When it emerges from the core (P waves) it refracts away from the normal to meet the surface at 140°. So from the earthquake to 105° around the Earth, P and S waves can be detected. From 140°, only P waves arrive. Between these angles there is a shadow zone. No waves reach the surface here. The size of the core can be found by finding these angles carefully and the times of waves reaching the surface. Scientists use data from lots of earthquakes. Earth acts like a magnet. This provides more evidence that the core is liquid.

Identifies parts of Earth and types of wave which can pass through them.

Correct scientific explanation.

Examiner summary

The explanations are correct. However, the answer is disjointed and could be clearer. Using longer sentences to connect ideas would help to make a level 3 answer. For example '… there is a shadow zone where no waves reach the surface.' The answer overall contains just enough detail to be awarded level 3. The magnetic field idea is worth considering but would be better with more detail to show its relevance.

Here are three student answers to the following question. Read the answers together with the examiner comments around and after them.

Question — Life of Stars — Grade D–A*

A cluster is a group of galaxies.

Scientists have discovered star-forming clouds in the space between the galaxies in the Virgo Cluster.

Our Sun is a star which formed from a cloud and will eventually become a white dwarf. Describe the sequence in the life of a star like our Sun from its formation in a cloud to its white dwarf stage. (6)

Student answer 1 — Extract typical of a level ① answer

More scientific detail could be given here, e.g. hydrogen gas and dust particles

> Stars form in clouds of gas which are many light years away. Then they become main seqeunce stars and burn up all their hidrogin. When this happens they blow up into a giant red star which fades to a white dwarf. Sometimes they eventually become a red or black dwarf.

Use correct scientific terms whenever possible: 'fuses' instead of 'burns up', 'red giant' instead of 'giant red' and 'expands' rather than 'blows up'.

Make sure spelling, punctuation and grammar is correct.

Examiner summary

This answer gives the stages more or less in the correct order. However, to improve the mark, the answer needs to include more details, e.g. the role of gravity in the formation of the star, the role of hydrogen gas in the nebula and the role of energy in an outwards push to prevent collapse of the star. Be sure to use technical words correctly and check spelling, punctuation and grammar.

Student answer 2 — Extract typical of a level ② answer

Good detail of the role of gravity here.

> The gas in a nebula comes together to make a star because of gravitational attraction. The main sequence star burns hydrogen as a fuel for millions of years. The heat prevents the star collapsing because the particles move and create a pressure which pushes outwards. When they finish fusing hydrogen they turn into red giants. After this they become a white dwarf and will eventually cool to become black dwarfs. If they are big stars much bigger than our Sun, they become a red supergiant instead and explode into a nova. Then cool down into a black hole.

The technical vocabulary is muddled here – the correct term (fusing) is used later on, but 'burn' is used here instead. Using 'fuse' you will show understanding that it is nuclear fusion in stars.

This is good physics but is not relevant to 'a star like our Sun' and will not gain credit here.

Examiner summary

This answer correctly gives the sequence of events in a star's life and includes some detail, e.g. the role of gravity in the 'birth' of the star. Most of the technical vocabulary is used correctly but it could also state that the gas is hydrogen. The answer has given some indication of why the main sequence star does not collapse. There is a good indication of time scale where millions of years are mentioned, but the answer could be improved by explaining the role of gravity and energy at the end of the star's life.

The role of gravity is stated clearly at star formation and with some idea throughout.

This shows understanding of the difference between main sequence and red giant stage.

Nebulas are clouds of dust and hydrogen gas. Gravity makes them collapse and heat up. They start to fuse hydrogen into helium and give out energy as light. This is a main sequence star. They stay like this for billions of years until all the hydrogen fuel is used up. They don't collapse because the particles are heated to a very high temperature which means they move very quickly. They try to spread out as they collide and push the rest of the star out. Also, the light and heat energy tries to escape and pushes out. This balances the inward pull of gravity. The star then becomes a red giant. Red giants fuse helium. After all the helium is used up the star becomes a white dwarf. Eventually the star cools so gravity pulls it into a black dwarf.

This shows the outward effects.

This shows why stars stay about the same size.

Examiner summary

This answer is very full and clear. There is enough detail at each stage, without writing too much unnecessary information which would use up time in the exam. The role of gravity at the birth and death of the star has been included and there is some explanation of why the star does not collapse under the effects of gravity. Appropriate technical language is used and spelled correctly, e.g. 'fuse' and 'red giant', and the answer is easy to read.

ResultsPlus

Move from level ① to level ②

- Make sure you know the stages of star development and can put them in the correct order.
- It is often a good idea to write each stage starting on a new line. That way if you have left something out or got them in the wrong order, you can number your sentences.
- Check through your answer for technical vocabulary. Have you used the correct terms in the correct way?
- Don't try to make your answer longer by copying information from the question.
- Try to offer some explanations for the movement from one stage of the sequence to the next.

Move from level ② to level ③

- Check that your answer is consistent and that it is clear what you refer back to.
- For this type of question where you have to give a sequence, you may find it useful to draw it as a flow diagram first.
- Think why the star moves from one stage to the next and why it stays more or less the same size in each stage.
- Plan your answer and check that you add as much detail as you can within your time limit. If you don't cross out the plan it will be marked and could gain you credit.

ResultsPlus

Read the passage and answer the questions that follow.

Is there life on another planet?

Many scientists believe that life exists on other planets. There is a vast number of stars in the Universe and many of these stars must have planets in orbit around them. The chances are that the right conditions for life exist on some of these planets.

- In what form might this life exist? Is it as simple microbes or are there other animals, similar to humans, out there?
- Could other life forms communicate with the Earth? The distances involved mean that it would take millions of years for signals to reach the Earth from another Solar System.
- Is it possible that life forms from another Solar System could be watching the Earth and even be on their way to visit us?

The right conditions for life to exist on a planet include a suitable composition of any atmosphere, a suitable temperature and the presence of water.

(i) These conditions have been measured on some planets in our Solar System, for example on Mars. Describe how scientists were able to do this. (3)

⚠️ **Correct answer:** Any two from this list, together with good, clear communication, would have got three marks:
- unmanned objects sent/landed (on Mars)
- sensed conditions/instruments used to measure conditions or gather data, e.g. temperature/atmosphere/photographs/composition of planet surface/collect sample material
- signals sent back to Earth

🔘 Some students spoilt their answer by referring to scientists on Mars. Nobody has been there. Robots have investigated and then sent information back to Earth. It is not enough to simply say 'scientists have used a computer to investigate'.

(ii) Explain why scientists cannot measure these conditions on planets in another galaxy. (3)

⚠️ **Correct answer:** You need to explain the following to get full marks:
- distance is too great/other galaxies are too far away
- it takes too long to get there/takes too long for signals to return
- technology not advanced enough

🔘 Try to make use of the information given in the question. The second bullet point refers to *millions* of years for the signal. Humans would of course take very much longer to travel.

 ResultsPlus
Build better answers

The diagram shows two types of brain wave.

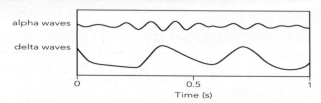

The frequency of alpha waves is about: (1)

A 0.1 Hz   **B** 1 Hz   **C** 10 Hz   **D** 100 Hz

⚠️ **Correct answer:** C

🔘 Some students forgot that frequency is the number of vibrations in each second. In the 1 s shown on the diagram, there are 10 alpha waves so the frequency is 10 Hz.

How many stars are in the Solar System? (1)

A none **B** one **C** thousands **D** millions

 Correct answer: B

As many as 80% of students chose '*millions*' as the answer to this question. Make sure you read the question carefully, so that you answer the question on the paper, not the one (maybe about a galaxy) that you think is there.

The table shows approximate travel times from Mars to three destinations.

	to reach the Earth	to reach the Sun	to reach a star
travel time for a radio signal from Mars	4 minutes	12 minutes	4 years
travel time for a space probe from Mars	6 months	18 months	X

Which of these will be correct for X? (1)

A 6 years **B** 60 years **C** thousands of years **D** space probes cannot leave our Solar System

Correct answer: C

You should make sure that you read the data carefully – almost half of students chose '6 years'. The important factor to consider here is the units of time. In the 'Earth' and 'Sun' columns, different units are used for the radio signal and for a space probe.

The diagrams show the line spectra from a nearby galaxy and a distant galaxy. Explain how the shift in the line spectrum for the distant galaxy supports the 'Big Bang' theory of the Universe. (2)

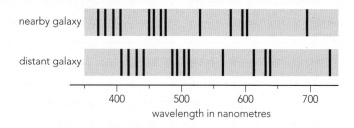

Correct answer: You need to include reference to the following to get the marks:
- the galaxies moving away from each other at different speeds • how this explains the expansion of the Universe

It is not enough just to say that the galaxies are moving. They must be moving relative to each other. Similarly, to generalise from this information about only two galaxies to the Universe as a whole, you should make clear that you are talking about more than just the two galaxies given in the data, Few people scored both marks.

Glossary

acid rain   Rain that is more acidic than normal, due to sulfur dioxides and nitrogen oxides dissolved in it.

acquired characteristic   A characteristic that is changed by the environment rather than inherited from your parents.

adaptation   Organisms have certain characteristics that allow them to survive in particular places. These characteristics are called adaptations.

addicted   When the body is dependent on a drug and doesn't work properly without it.

ⓗ algae   A group of organisms that can make food using photosynthesis but do not have leaves or roots, like plants. These organisms include seaweeds and some single-celled organisms (singular: alga).

alkali   A soluble base.

alkane   A hydrocarbon with only single bonds between two carbon atoms.

alkene   A hydrocarbon with at least one double bond between carbon atoms.

allele   Every gene comes in different types called alleles. So a gene for eye colour may come in a 'blue type' allele and a 'brown type' allele.

alloy   Mixture of metals.

alloy steel   An alloy of iron with other metals (steel itself is not an alloy).

alpha particles   A type of ionising radiation emitted by a radioactive material.

alternating current   Current whose direction changes many times each second.

amphibians   Vertebrates that have moist, permeable skin and lay jelly-coated eggs in water.

amplitude   The maximum distance of particles in a wave from their normal positions.

antacid   A compound that is used to neutralise acid in the stomach.

antibacterial   Antibiotic that kills or slows the growth of bacteria.

antibiotic   Substance that can kill or slow the growth of microorganisms.

antifungal   Antibiotic that kills or slows the growth of fungi.

antiseptic   Substance that is used to stop the spread of pathogens.

ⓗ aqueous solution   Mixture formed when a substance dissolves in water.

atmosphere   Layer of gases that surrounds the Earth.

atom   The smallest part of an element that can take part in chemical reactions.

autotrophic feeding   Make food from small molecules using an energy source, such as light in photosynthesis.

auxin   Plant hormone that affects the growth and elongation of cells in plants.

axon   The long extension of a neurone that carries an impulse away from the cell body towards other neurones

bacteria   Simple organisms consisting of one cell that does not contain a nucleus (singular: bacterium).

bacterium   Microscopic organism, some kinds of which are pathogens, such as Salmonella (plural: bacteria).

ⓗ balanced equation   Description of a reaction using the symbols and formulae of the reactants and products.

base   A substance that will react with an acid to form only salt and water.

beta particles   A type of ionising radiation emitted by a radioactive material.

Big Bang theory   The theory that says that the Universe began from a tiny point with huge energy, and has been expanding ever since.

binomial system   System of naming organisms using two Latin words.

biodegradable   A substance that can be broken down by microorganisms.

biodiesel   Diesel fuel made from plant material.

biodiversity   A variety of species of plants and animals.

biofuel   Any fuel made by humans from animal or plant materials that have recently died.

biogas   Methane made by from animal manure or biodegradable waste from homes and farms.

biomass   The total mass in living organisms, usually shown as the mass after drying.

birds   Vertebrates that have lungs, feathers and beaks, and lay hard-shelled eggs.

bitumen   The fraction of crude oil with the longest molecules. It is used for making roads and roofs.

ⓗ black hole   Core of a red supergiant that has collapsed. Black holes are formed if the remaining core has a mass more than three or four times the mass of the Sun.

blackspot fungus   Fungus that damages roses and is killed by sulfur dioxide in the air.

blood glucose regulation   The control of the concentration of glucose in the blood by the body.

blood worm   Aquatic species that is an indicator of polluted water.

Body Mass Index (BMI)   Estimate of how healthy a person's mass is for their height.

boiling point   The temperature at which all of a substance starts to turn into a gas.

bonds   Forces of attraction holding atoms together in a molecule.

bromine test   A test for unsaturation. Bromine water turns colourless when mixed with an unsaturated molecule.

bromine water   A solution of bromine in water that turns colourless when mixed with an unsaturated molecule.

ⓗ carats   A measure of the purity of gold with pure gold being 24 carat.

carbon brushes   Blocks of carbon connected to a circuit. They press against slip rings to make an electrical connection between the circuit and the spinning coil in a generator.

carbon cycle   A sequence of processes by which carbon moves from the atmosphere, through living and dead organisms, into sediments and into the atmosphere again.

carbon monoxide   A toxic gas (CO), it is found in tobacco smoke that replaces oxygen in the blood, and so reduces the amount of oxygen carried around the body.

carbon neutral   A fuel or process that does not add any carbon dioxide to the atmosphere overall.

carcinogen   Something that causes cancer, such as tar in tobacco smoke.

cataracts   When the lens of the eye becomes cloudy.

cell membrane   Thin layer around a cell that controls what goes into and out of a cell.

cell wall   Outer stiff part of some cells that helps to support a cell. Plant cell walls are made out of cellulose.

cement   Material made by heating limestone with clay.

central nervous system (CNS)   The spinal cord and brain. These two organs form the main part of the nervous system, processing and controlling the transmission of electrical impulses.

chalk   Sedimentary rock, mostly calcium carbonate, made from the remains of marine micro-organisms.

characteristics   The features of an organism.

chemical defence   Preventing attack by the use of chemicals. For example, many plants produce chemicals that taste unpleasant and this puts herbivores off eating them.

chemical formula   A combination of symbols and numbers that shows how many atoms of different kinds are in a particular molecule. In compounds that do not form molecules, it shows the ratio of elements in the compound.

chemical potential energy   The energy stored in matter, which can be released by a chemical reaction (such as combustion).

Ⓗ **chemosynthetic bacteria**   Bacteria that get the energy they need to make their food from breaking down chemicals; they are producers.

chloroplast   Green disc found in plant cells, which is used to make food for the plant using photosynthesis.

Chordata   Animals that have a supporting rod along the length of their body. All vertebrates belong to this group. (Also known as chordates).

chromosome   A long thread of a molecule called DNA. Each chromosome contains a series of genes along its length.

cirrhosis   Damage to the liver caused by drinking large amounts of alcohol over a long time.

classification   The process of sorting organisms into groups based on their characteristics.

cleaner fish   Fish that dead skin and parasites from the skin of other fish.

climate change   Changes to the Earth's climate, or changes in weather patterns on a global scale.

clinically obese   When someone is very overweight for their height to an extent that has been shown by doctors to cause health problems. For adults this is defined as having a BMI of over 30.

CNS   Abbreviation for central nervous system.

combustion   Chemical reaction when substances burn, combining with oxygen to produce heat and waste products such as carbon dioxide.

competition   When organisms need the same resources as each other, they struggle against each other to get those resources.

complete combustion   Combustion of hydrocarbons with enough oxygen present to convert all the fuel into carbon dioxide and water.

compound   A substance containing two or more elements chemically joined together.

concentration   The amount of a substance dissolved in a certain quantity of liquid (e.g. g/cm^3, $mg/100\ cm^3$).

concrete   Material made by mixing cement with sand, stones and water.

conductor, electrical   A sample which allows electricity to pass through it.

consumer   An animal, because it consumes (eats) other organisms.

continuous variation   When a variable can have any numerical value. Human height is a continuous variable.

control   An experiment that uses the same apparatus as the main part of the experiment but the variable that is being changed is not applied.

convection current   A current caused by parts of a fluid being at a different temperature and so a different density to the rest of the fluid.

convention   A standard way of doing something.

converge   To come together.

convex lens   A lens which is thicker in the middle than at its edges.

core   The middle of the Earth.

corrosion   Occurs when a metal is converted to its oxide by the action of moist air.

cosmic microwave background radiation (CMB)   Microwave radiation received from all over the sky, originating at the Big Bang.

cost-efficient   Something that saves a lot of money compared to how much it costs.

cracking   Splitting up long hydrocarbon molecules into shorter ones.

criterion   Rule or test that can be used to judge how to make a fair decision (plural: criteria).

crude oil   A mixture of hydrocarbons formed from dead microscopic organisms by heat and pressure over millions of years.

crust   The outer surface of the Earth.

crystal   A solid in which the particles are arranged in an ordered pattern.

current   A flow of charge.

Ⓗ **cutting**   Part of a plant (normally a leaf or stem) from which a new plant can be grown.

cystic fibrosis (CF)   A genetic disorder caused by inheriting two copies of a recessive allele. It causes thick mucus to collect in the lungs, making breathing difficult, and also stops food being digested very well.

cytoplasm   Jelly-like part inside a cell where the cell's activities happen.

decay Process in which complex substances in dead plant and animal biomass are broken down by decomposers into simpler substances.

decommission To dismantle safely.

decompose To break down larger molecules into smaller ones.

decomposer Organism that feeds on dead material, causing decay.

deforestation The destruction of the world's trees and forests.

dendrite Many fine extensions of a dendron of a neurone that collect impulses from other neurones or receptors.

dendron Extension of a neurone that carries the impulse to the cell body. It ends in many dendrites and is usually shorter than an axon.

denitrifying bacteria Bacteria that break down more complex nitrogen compounds into simpler ones, such as nitrates to nitrites, or nitrites to ammonia.

dense Something which has a lot of mass in a given volume.

density The mass of a substance per unit volume; the unit is usually g/cm^3.

depressant Substance that slows down the activity of neurones in the brain.

dermis The layer of skin below the surface epidermis, where nerve endings, blood vessels and other structures are found.

diabetes Disease in which the body cannot control blood glucose concentration at the correct level.

diesel oil Fraction of crude oil used as a fuel in lorries and some cars.

digest Break down, as when our digestive system breaks up food into simpler substances.

direct current (d.c.) A current that flows in one direction only, such as the current produced by a cell.

discontinuous variation When a variable cannot have a continuous range of options, for example days of the week, shoe sizes.

DNA Deoxyribose nucleic acid. Chemical that makes up genes and chromosomes; the instructions for a cell's growth and activity.

dominant allele Version of a gene (allele) that will always have an effect (as opposed to a recessive allele, whose effect will not be seen if a dominant allele is present).

donor Person who gives, such as the person who gives an organ for transplant into another person.

Doppler effect The change in the pitch of a sound coming from a moving source.

double bond When two bonds join a pair of the same atoms (usually two carbon atoms).

drug A substance that changes the way the body works.

ductile Can be stretched into wires.

dynamo A small machine that generates electricity using a spinning magnet.

earthquake A sudden movement within the Earth, releasing a large amount of stored energy .

effector Muscle or gland in the body that performs and action when an impulse from the nervous system is received.

efficiency The proportion of input energy that is transferred to a useful form.

egg cell Another term for ovum.

elastic potential energy The energy stored by things that have been stretched or squashed and can spring back .

electrical energy Energy transferred by electricity.

electricity A flow of charged particles that can be used to transfer energy.

electrode Part of an electrical circuit that makes contact with a sample.

electrolysis Passing a direct electrical current through an electrolyte to decompose it.

electrolyte A liquid which conducts the electric current and is decomposed by it.

electromagnet A magnet made using a coil of wire with electricity flowing through it.

electromagnetic induction Process that creates a current in a wire when the wire is moved relative to a magnetic field, or when the magnetic field around it changes.

electromagnetic radiation A form of energy transfer, including radio waves, microwaves, infrared, visible light, ultraviolet, X-rays and gamma rays.

electromagnetic spectrum The entire frequency range of electromagnetic waves.

electromagnetic waves A group of waves that all travel at the same speed in a vacuum, and are all transverse.

element Substance that cannot be split up into simpler substances.

endocrine gland A gland that makes and releases hormones.

energy transfer Energy being moved from one place to another, possibly with a change in the form of energy at the same time.

environmental variation Differences between the characteristics of organisms caused by their environment.

epicentre The point on the surface of the Earth directly above the focus of an earthquake.

erector muscle Muscle in the dermis of the skin that raises a body hair.

erosion Removal or transportation of broken up rocks.

ethanol (alcohol) A fuel made by processing sugar beet or sugar cane.

ethical decision Decision about what is right or wrong.

eutrophication The addition of chemicals to water, such as nitrates and phosphates, which encourages plant growth.

evolution Gradual change over a period of time.

external fertilisation When fertilisation happens outside the body of an animal.

extinction The dying out of a species so that it no longer exists.

extraction (metal) Reaction in which a metal is produced from a compound in an ore.

extraterrestrials   Beings that live on planets other than the Earth.

eyepiece lens   The lens on a telescope closest to the observer's eye.

faeces   Waste material from food which was not absorbed by the body.

family   A classification group that contains genera with similar characteristics.

family pedigree chart   A chart showing which members of a family suffer from a genetic disorder.

feeding relationship   The relationship between two organisms where one eats the other.

fertile   Able to produce offspring.

fertiliser   Chemical compounds added by farmers to soil to increase the rate of growth of crops.

(H) fineness   A measure of the purity of gold in parts per thousand with pure gold having a fineness of 1000.

fish   Vertebrates that have wet scales, gills and lay jelly-coated eggs.

fluorescence   Absorbing radiation of one wavelength and re-emitting the energy at a different wavelength (usually so that it becomes visible).

fluorescent lamp   Lamp which works by fluorescence. It produces ultraviolet radiation, which is then absorbed by a coating inside the bulb and re-emitted as visible light.

focal length   The distance from a convex lens to the point where the image of a very distant object is produced.

focus   The place where an earthquake begins, usually under the surface.

food chain   A diagram that shows with arrows the flow of food and energy from a producer to the animal that eats that producer, the animal that eats that animal, and so on.

food web   A diagram of interlinked food chains that show how the feeding relationships in one habitat are interdependent.

formula (chemical)   The composition of a substance showing the symbols of the elements it contains, and the ratios in which their atoms are present.

fossil   The remains or trace evidence of prehistoric living organisms.

fossil fuel   Non-renewable fuels such as coal, oil and natural gas that have formed over millions of years from dead plants and animals.

fractional distillation   The process by which a mixture of two or more liquids is separated, for example crude oil is separated into different fractions.

fractions   The different mixtures into which crude oil is separated.

frequency   The number of waves passing a point each second.

freshwater shrimp   Aquatic species that is an indicator of clean water.

fuel cell   A device which produces electricity from hydrogen and oxygen without burning.

fuel oil   Crude oil fraction used as fuel for ships and for lubrication.

fungus   Organism that feeds on dead or decaying material, such as athlete's foot fungus (plural: fungi).

fusion reaction   When the nuclei of two atoms join together and release energy.

galaxy   A group of millions of stars held together by gravity.

gamete   Cell that contains only half the normal number of chromosomes. It joins with another sex cell during fertilisation, to produce a fertilised egg cell, which contains a full set of chromosomes.

gamma rays   High frequency electromagnetic radiation.

gene   A section of DNA that carries the instructions for a characteristic.

generator   A machine that makes electricity when it turns.

genetic cross diagram   Diagram showing how the alleles in two parents may form different combinations in the offspring, when the parents reproduce.

genetic disorder   A disease caused by alleles.

genetic variation   Variation in characteristics caused by the instructions within cells.

genotype   The alleles for a certain characteristic that are found in an organism.

genus   A classification group that contains species with similar characteristics (plural genera).

geocentric   Earth-centred.

geothermal energy   Energy transferred from hot rocks deep beneath the Earth's surface.

geotropism   Tropism in response to gravity.

germinate   When a seed starts to grow a shoot and a root.

gibberellins   Set of plant hormones that can cause seeds to germinate, and flowers and fruits to form.

gland   A part of the body the makes substances and then releases them.

global warming   The increase in the Earth's average temperature likely to be caused by increased amounts of carbon dioxide in the atmosphere.

glucagon   Hormone released by endocrine glands in the pancreas, which increases the blood glucose concentration by causing cells, especially those in the liver, to turn glucose into glycogen.

glucose   A sugar, which is produced by the digestion of carbohydrates and is needed be cells for respiration.

glycogen   A storage material made from glucose.

granite   Igneous rock with relatively large crystals, formed when magma cools and solidifies slowly.

gravitational potential energy   The energy stored in things that can fall.

greenhouse effect   When gases in the atmosphere trap heat energy and keep the Earth warm.

greenhouse gases   Gases that help to trap heat in the atmosphere. They include carbon dioxide, methane and water vapour.

habitat   The place an organism lives in, for example woodland.

hallucinogen   Substance that distorts sense perception.

hazard symbol   A symbol used on containers that warns people about the dangers of the contents.

hazardous   Something that could be harmful or dangerous if not used and stored properly.

heliocentric   Sun-centred.

hertz (Hz)   The unit for frequency, 1 hertz is 1 wave per second.

heterotrophic feeding   Getting food by eating and digesting the tissues of other organisms.

heterozygous   If both alleles for a characteristic are different, the organism is heterozygous for that characteristic.

homeostasis   Controlling the internal environment of the body at stable levels.

homeotherm   An animal that keeps its body temperature more constant than the surroundings, and often warmer, by releasing heat from reactions in its body.

homozygous   If both alleles for a characteristic are the same, the organism is homozygous for that characteristic.

hormone   A substance that is made and released in one part of the body and that has an effect on another part of the body (a chemical messenger).

host   Organism that provides food for a parasite.

hybrid   An organism that is the result of breeding together two different species. A hybrid has characteristics from each species.

hydrocarbon   A compound containing only hydrogen and carbon atoms.

hydroelectricity   Electricity generated using the transfer of gravitational potential energy to kinetic energy by falling water.

hydrothermal vent   Area on the seabed where hot gases and water are forced up from below, by being heated by magma below the sea floor.

hypothalamus   Part of the brain that controls body temperature.

igneous rock   Rock formed when magma cools and solidifies.

ignite   To start burning.

illegal   Against the law, often punishable by a fine or imprisonment.

illuminate   To light up with visible light.

image   Picture formed by a mirror or lens.

impulse   Electrical signal transmitted along a neurone.

incinerate   To burn.

incomplete combustion   Combustion that occurs without enough oxygen to completely oxidise all the fuel. Incomplete combustion of hydrocarbons produces carbon dioxide, carbon monoxide and carbon (soot).

indicator   A solution that changes colour in acid and alkaline solutions.

indicator species   Species that is particularly sensitive to or tolerant of pollution, so that its presence or absence can be used as a measure of the pollution.

induced current   The current that flows in a wire that is moving relative to a magnetic field.

infectious disease   Illness that is caused by a microorganism and can be caught from an infected person.

infrared (IR)   Electromagnetic radiation that we can feel as heat.

infrasound   Sound waves with a frequency below 20 Hz, which is too low for the human ear to detect.

inherited variation   Variation caused by genes.

inhibition   The feeling that you can't or shouldn't do something.

insoluble   Substance which does not dissolve in a given solvent.

insulin   Hormone released by endocrine glands in the pancreas, which decreases the blood glucose concentration by causing cells, especially those in the liver, to turn glycogen into glucose.

interbreed   Reproduce with other members of the same group.

interdependent   Depending on each other.

interface   Junction between two different materials.

internal environment   The conditions inside your body.

internal fertilisation   When fertilisation happens inside the body of an animal.

invertebrate   Animal with no backbone.

ion   An atom that has become electrically charged.

ionising radiation   Radiation or certain types of high-energy particles that can cause atoms to become electrically charged (to become ions).

IR   Abbreviation for infrared.

iron seeding   Adding iron compounds to the oceans to encourage organisms to grow and remove carbon from the carbon cycle.

joule (J)   The unit of energy.

kerosene   Crude oil fraction used as fuel for jet engines.

key   Diagram containing a set of questions or statements that can be used to work out the name of an organism.

kingdom   The largest division in the classification of organisms.

Kingdom Animalia   A large group of complex organisms that have nervous systems.

Kingdom Fungi   A large group of organisms that cannot make their own food. They reproduce using spores (rather than seeds) and live attached to their food sources. They include moulds, yeasts, mushrooms and toadstools.

Kingdom Plantae   A large group of organisms that usually have cells containing chloroplasts and can make their own food, using photosynthesis.

Kingdom Prokaryotae   A large group of organisms that consist of one cell, which does not have a nucleus. Bacteria are an example.

Kingdom Protoctista   A large group of organisms that do not fit into any of the other four kingdoms. Algae are an example.

kidney   An organ that is important in removing extra water and salts from the blood, by producing urine.

kilowatt (kW)   A unit for measuring power. 1 kW = 1000 W.

kilowatt-hour (kWh)   The amount of energy transferred in an hour by a 1 kW appliance.

kinetic energy   Movement energy.

lander   A space vehicle that lands on a planet or moon.

lava   Molten rock on the Earth's surface.

law of conservation of energy   States that energy cannot be created or destroyed, although energy may transform from one form into another.

legal   Allowed by law.

(H) legume   Plant of the pea family, including peas and beans.

lens   shaped piece of glass or other transparent material that refracts light in particular ways.

lichen   Mutualistic relationship between a fungus and an alga; different species of lichen are affected differently by air pollution, so they can be used as pollution indicators.

light energy   The energy of visible light.

light waves   Electromagnetic waves that can be detected by the human eye.

limestone   Sedimentary rock, mostly calcium carbonate, made from the remains of marine organisms.

limewater   A solution of calcium hydroxide that turns milky when carbon dioxide is bubbled through it.

litmus paper   Blue litmus paper turns red in acid solutions and red litmus paper turns blue in alkaline solutions.

longitudinal   E.g. sound waves, where the direction of energy transfer is parallel to the direction of vibration which causes them.

longitudinal waves   Waves in which particles move back and forth in parallel with the direction of movement of the energy.

magma   Molten rock inside the Earth's crust.

magnetic field   The area around a magnet where it can affect magnetic materials or induce a current.

magnification   The number of times larger an image is than the object that produced it.

magnify   Enlarge or make things look bigger.

main sequence star   A star during the main part of its life cycle, where it is using hydrogen fuel.

(H) malaria   A dangerous disease caused by a protist that causes serious fever, headaches and vomiting and can lead to death.

malleable   Can be hammered into shape.

mammals   Vertebrates that have fur, lungs and produce milk on which to feed their young.

mantle   The part of the Earth between the crust and the core.

marble   Metamorphic rock, mostly calcium carbonate, formed from limestone or chalk.

medium   Something through which waves travel.

metamorphic rock   Rock formed by the action of very high temperature and pressure on other rocks.

microwaves   A type of electromagnetic wave.

Milky Way   The name of our galaxy.

mixture   A substance containing two or more different substances that are not joined together.

model   An example of something happening which explains how a scientific idea should be understood.

molecule   Two or more atoms joined together.

monomer   Small molecule used to make a polymer.

motor neurone   Neurone that carries impulses to effectors.

(H) MRSA   Methicillin-resistant Staphylococcus aureus, a strain of bacterium that is resistant to many kinds of antibiotics.

multicellular   Made of more than one cell.

mutation   A change in the DNA of a gene.

mutualism   A relationship between organisms where both benefit.

myelin sheath   Fatty covering around the axons of many neurones. It speeds up the transmission of impulses along their length and helps to insulate them from one another.

naked eye   Obsevation made using just the eyes, without using a telescope or any other aid.

narcotic   A drug that makes you feel sleepy.

National Grid   The system of wires and transformers that distributes electricity around the country.

natural selection   A process in which the organisms that are best suited to the conditions in their habitats are more likely to survive.

nebula   A cloud of gas in space. Some objects that look like nebulae are actually clusters of stars or other galaxies, (plural: nebulae).

negative feedback   A control mechanism that reacts to a change in a condition (such as temperature) by trying to bring the condition back to a normal level.

negative powers   A power is written in the form a^n, where n shows the number of times that a is multiplied by itself. A negative power is written in the form of a^{-n} and is the inverse, which we can write as $1/a^n$.

negative tropism   Tropism in which the response is away from the stimulus.

nerve   Bundle of neurones.

nerve cell   Another term for neurone.

nervous system   An organ system that includes the brain and nerves, which carries information around an organism.

neurone   A cell that transmits electrical impulses in the nervous system.

neurotransmission   Impulses passing from neurone to neurone.

neurotransmitter   Substance that diffuses across the gap between two neurones at a synapse, and triggers an impulse to be generated in the neurone on the other side of the synapse.

neutralisation reaction   Reaction in which a base or alkali reacts with an acid.

(H) neutron star   Core of a red supergiant that has collapsed.

nicotine   Stimulant in tobacco smoke which is addictive and makes it difficult to give up smoking.

nitinol   An alloy of nickel and titanium, which is a shape memory alloy.

Glossary

nitrifying bacteria Bacteria that make more complex nitrogen compounds from simpler ones, such as nitrates from nitrites, or nitrites from ammonia.

nitrogen cycle A sequence of processes by which nitrogen moves from the atmosphere through living and dead organisms, into the soil and back to the atmosphere.

nitrogen-fixing bacteria Bacteria that can take nitrogen from the atmosphere and convert it to more complex nitrogen compounds, such as ammonia.

noble gases Unreactive gases in Group 0 of the Periodic Table.

non-renewable Any energy resource that will run out one day.

non-renewable resources Resources that cannot be replaced once they have been used. Non-renewable resources will eventually run out.

normal Line at right angles to the surface of a mirror or lens where a ray of light hits it.

normal distribution curve A graph of variation in a characteristic for a population, with a bell-shaped curve that shows most values in the middle of the range and a few extreme values.

nuclear potential energy The energy stored in the nuclei of atoms.

nuclear power Generating electricity using energy stored in nuclear fuels.

nucleus The central part of an atom, containing protons and neutrons. The control centre of a cell (plural: nuclei).

objective lens The lens in a telescope nearest the object to be observed.

optic fibre Glass or plastic strand which can direct light or infrared radiation.

orbit The path taken by a planet around the Sun, or a satellite around a planet.

ore Rock from which a metal can be extracted for profit.

osmoregulation Controlling the amount of water in the body.

oviparous Offspring develop in eggs, as in birds.

ovum The female gamete in plants and animals (plural: ova).

oxidation Occurs when oxygen is added to an element or compound.

P waves Longitudinal seismic waves that travel through the Earth.

painkiller Substance that blocks the transmission of pain responses via neurones to the brain.

pancreas Organ in the body that produces some digestive enzymes as well as insulin and glucagon.

parasite Organism that lives on or in a host organism and takes food from it while it is alive.

parasitism A feeding relationship where one organism benefits and the other is harmed.

patent An order made by a government that means that only the creator of an invention is allowed to make or sell that invention.

pathogen Microorganism that causes disease.

payback time The time it takes to get back in savings the money spent on making a change.

pedigree analysis When doctors study family pedigree charts to assess the probability that a couple may have passed on a genetic disorder to their child.

permeable A solid that allows a liquid or a gas to pass through it is described as being permeable.

petrol Crude oil fraction used as fuel for cars.

pH scale A scale from 1 to 14 showing acidity or alkalinity. Numbers below 7 are acids. Numbers above 7 are alkalis. pH 7 is neutral.

phenotype The characteristics that a certain set of alleles cause.

photosynthesis Set of chemical reactions in plants that allow them to produce their own food (glucose) using water and carbon dioxide and releasing oxygen as a waste product. The process is powered by light from the Sun.

phototropism Tropism in response to light.

physical barrier A structure that stops something from entering a certain area. For example, the body has physical barriers like the skin, which stop microbes from getting inside the body.

pitch Whether a sound is low or high.

plant growth substances A substance released by a part of a plant that has an effect on the cells of that part or another part of the plant, usually causing the cells to grow or develop in a different manner. Another term for this is plant hormone.

plastic The common name for many polymers.

plate Sections of the outermost layer of the Earth that can move relative to each other.

poikilotherm An animal whose body temperature varies with the temperature of the environment around it.

pollen grains The male gamete in plants.

pollutant A substance that harms living organisms when released into the environment, often waste products of human activity.

poly(chloroethene) A polymer made from chloroethene monomers. Also known as PVC.

poly(ethene) A polymer made from ethene monomers.

poly(propene) A polymer made from propene monomers.

poly(tetrafluoroethene) The chemical name for PTFE, which is also known as Teflon®.

polymer A long molecule made by joining many smaller molecules (monomers) together.

polymerisation The process of making a polymer.

population growth Increase in population size over time.

positive gravitropism Tropism in which the response is towards the stimulus.

power How quickly something transfers energy.

precipitate Insoluble product formed in a precipitation reaction.

precipitation reaction Reaction in which an insoluble product is formed from soluble reactants.

predator Animal that kills other animals to eat.

prey An organism that is hunted and killed by a predator.

primary coil The coil on a transformer to which the electricity supply is connected.

primary consumer An animal that eats producers (i.e. a herbivore).

primary data Data that you collect yourself.

primary mirror The main mirror in a reflecting telescope which converges light to form an image.

(H) probability The likelihood of something happening, often shown as a percentage chance. For example, there is a 50% chance that it will rain tomorrow.

producer Organism that makes its own food, such as a plant using photosynthesis.

product Substance formed by a chemical reaction.

protoctists Simple organisms belonging to the kingdom Protoctista. Most protoctists are unicellular, although some (such as seaweeds) are multicellular. They have complex cells with a nucleus.

protostar A cloud of gas drawn together by gravity that has not yet started to produce its own energy.

protozoan Type of one-celled protoctist that requires a source of food (i.e. it cannot photosynthesise as some other protoctists can).

PTFE The abbreviation for poly(tetrachloroethene), also known as Teflon®.

Punnett square Diagram used to predict the different characteristics that will be present in the offspring of two organisms with known combinations of alleles. You can use the square to work out the probability (how likely it is) that offspring will inherit a certain feature.

PVC Another name for poly(chloroethene).

pyramid of biomass Diagram showing the biomass in each trophic level of a food chain.

quarry Shallow mine on the Earth's surface from where rocks are removed.

radio waves A part of the electromagnetic spectrum.

radioactive Any material that gives out alpha, beta or gamma radiation.

radiotherapy Cancer treatment in which a patient is given gamma radiation to kill the cancer cells.

rate How quickly something happens.

reactant Substance used up in a chemical reaction.

reaction time How long it takes to respond to a stimulus.

(H) reactivity series A list of metals arranged in order of reactivity.

receptor cell Cell that receives a stimulus and converts it into an electrical impulse to be sent to the brain and/or spinal cord.

recessive allele Version of a gene (allele) that will only have an effect if the other allele is also recessive.

recycle To extract a material from waste so that it can be reused or processed to make a new object or material.

recycling Taking materials from waste and making them into useful products again.

red giant A star that has used up all the hydrogen in its core and is now using helium as a fuel. It is bigger than a normal star.

(H) red supergiant A star that has used up all the hydrogen in its core and has a mass much higher than the Sun.

red-shift Waves emitted by something moving away from an observer have their wavelength increased and frequency decreased compared to waves from a stationary object.

reduction Occurs when oxygen is removed from a compound.

reflect When a wave bounces off a boundary between two materials.

reflecting telescope A telescope in which the focussing of the main image is done by a curved mirror.

(H) reflex Response to a stimulus that does not require processing by the brain. The response is automatic.

(H) reflex arc Connection of a sensory neurone to a motor neurone (often via a relay neurone) that allows reflex actions to occur.

refracting telescope A telescope consisting of a series of lenses.

refraction The change of speed and direction of a wave when it enters a new material.

relay neurone A short type of neurone, found in the spinal cord and brain, that link with sensory, motor and other relay neurones.

renewable Any energy resource that will not run out.

renewable energy resources Resources that will not run out, such as solar or wind energy.

reptiles Vertebrates that have lungs, dry and scaly skin, and lay leathery-shelled eggs.

resistant An organism that has evolved so it is not affected by substances that would usually kill it. In the case of bacteria, no longer killed by an antibiotic; in the case of rats, no longer killed by warfarin poison. Organisms may vary in the range of their resistance, so that although some organisms may be killed, others may be made very ill but recover, while a few may be unaffected.

respiration Process that takes in oxygen and releases carbon dioxide, which living organisms use to release energy from food for all their activities.

response Action that occurs due to a stimulus.

ring species A ring of populations, in which neighbouring populations that can interbreed but the populations at the the two ends of the chain cannot (despite the fact that they may both live in the same area).

root nodule Small structure that legume plants make on their roots for nitrogen-fixing bacteria to live in.

(H) rooting powder Powder that contains plant hormones called auxins that help plant cuttings to grow roots quickly.

rover A space vehicle that can move around on a planet or moon.

rusting The corrosion of iron.

S waves Transverse seismic waves that travel through the Earth.

salt A compound formed by neutralisation of an acid by a base. The first part of the name comes from the metal in the metal oxide, hydroxide or carbonate. The second part of the name comes from the acid.

saprophytic feeding Getting food by digesting the tissues of other organisms outside the body and absorbing the digested food.

saturated A molecule with only single bonds between the carbon atoms.

Search for Extraterrestrial Intelligence (SETI) Looking for intelligent life beyond the Earth by trying to detect radio signals from them.

sebaceous glands Glands at the base of skin hairs that release oil onto the skin surface, keeping the skin lubricated and healthy.

secondary coil The coil on a transformer where the changed voltage is obtained.

secondary consumer An animal that eats primary consumers.

secondary data Data that you use from someone else's report, book, TV programme etc.

sediment Process by which solid material that settles to the bottom of a fluid, such as water.

sedimentary rock Rock made by compression of layers of solid material (sediment) that settles to the bottom of a fluid, such as water.

seismic waves Waves produced by an explosion or earthquake and which travel through the Earth. They include P-waves and S-waves.

seismometer An instrument that detects seismic waves.

H selective weedkiller Weedkiller that contains artificial plant hormones and will kill only certain types of plants. Most selective weedkillers kill plants with broad leaves and not those with narrow leaves.

sense organ Organ that contains receptor cells.

sensory neurone Neurone that carries impulses from receptors.

sex cells Another term for gamete.

H shape memory alloy An alloy that returns to its original shape with a change in conditions, often temperature.

sickle cell disease A genetic disorder caused by inheriting two copies of a recessive allele. It causes tiredness, shortness of breath and periods of extreme pain in the joints.

skin cancer A cancer or cancerous tumour on the skin.

slip rings Rings connected to the coil of a generator to make an electrical connection to a circuit. They allow the coil to spin without the wires becoming twisted.

sludgeworm Aquatic species that is an indicator of polluted water.

H smart material A material with a property that changes with a change in conditions.

solar cells A device that converts light energy into electrical energy.

solar energy Energy from the Sun.

Solar System An area of space in which objects are influenced by the Sun's gravity.

solidify To become solid.

soluble Substance which dissolves in a given solvent.

sonar A way of finding the distance to an underwater object (such as the sea floor) by timing how long it takes for a pulse of ultrasound to be reflected.

soot Tiny particles of solid carbon produced by incomplete combustion.

sound energy The energy transferred by sound waves.

sound wave Transverse seismic waves that travel through the earth.

space probe A space vehicle that can be put into orbit around a planet or moon, or parachuted down through the atmosphere.

speciation Formation of new species, such as when populations of a species are separated geographically and evolve until they are no longer capable of interbreeding.

species Each different type of organism is called a species. The members of a species can reproduce with each other to produce offspring that will also be able to reproduce.

spectrometer An instrument that can split up light to show the colours of the spectrum.

spectrum The range of colours between red and violet obtained when white light is split up using a prism.

sperm cells The male gamete in animals.

spinal cord Large bundle of nerves, leading from the brain and down the back.

stable Something that stays the same, without changing.

star A large ball of gas that produces heat and light energy from fusion reactions.

H state symbol Letter or letters to show the state of a substance.

Steady State theory The theory that the Universe is expanding but new matter is continually being created, so the Universe will always appear the same.

step-down transformer A transformer that reduces the voltage.

step-up transformer A transformer that increases the voltage.

sterilise To kill microorganisms such as bacteria.

stimulant Substance that increases the speed of transmission of nerve impulses across synapses.

stimulus Change in an environmental factor that is detected by receptors, (plural: stimuli).

stonefly larva Aquatic species that is an indicator of clean water.

subcutaneous fat Layer of fat under the skin.

H supernova An explosion produced when the core of a red supergiant collapses (plural: supernovae).

survival of the fittest See 'natural selection'.

sweat gland A gland found in the skin that produces sweat.

synapse Point at which two neurones meet. There is a tiny gap between neurones at a synapse, which cannot transmit an electrical impulse.

system Anything in which we are studying changes. Systems can be very complex (e.g. a planet) or quite simple (e.g. a piece of metal).

tar Sticky black substance in tobacco smoke that contains carcinogens.

target organ   An organ on which a hormone has an effect.

tectonic plate   Pieces of the surface of the Earth, which can move around very slowly.

Teflon®   The brand name for PTFE.

telescope   A device for producing magnified images of distant objects.

thermal decomposition   Reaction in which one substance breaks down when heated to form two or more new substances.

thermal energy   Energy transferred by heating.

thermal imaging   Photography that uses a detector of infrared radiation.

thermoregulation   The control of temperature inside the body by mechanisms in the body.

tidal power   Generating electricity using the movement of the tides.

trace   A small amount of.

transfer   To move something.

transformer   A device consisting of two coils of wire on an iron core which can change the voltage of an alternating electricity supply.

transplant   Taking an organ from one person and putting it into the body of another (often to save the life of that other person).

transverse waves   Waves in which movements are at right angles to the direction of movement of the energy.

trophic level   One level of a food chain, such as producer, herbivore, carnivore.

tropism   A response to a stimulus in which an organism grows towards or away from the stimulus.

tsunami   A huge wave caused by an earthquake or landslide on the sea bed.

Type 1 diabetes   Type of diabetes in which the pancreas does not produce insulin.

Type 2 diabetes   Type of diabetes in which cells, especially those in the liver, do not respond to insulin.

ultrasound   Sound waves with a frequency above 20 000 Hz, which is too high for the human ear to detect.

ultrasound scan   A way of making an image of part of the body (usually a foetus) using ultrasound waves reflected from parts of the inside of the body.

ultraviolet (UV)   Electromagnetic radiation that has a shorter wavelength than visible light but longer than X-rays.

unicellular   Made of one cell.

Unit   The unit of measurement for the amount of energy transferred. A Unit is the same as a kilowatt-hour.

universal indicator   A mixture of different indicators giving a different colour at different points on the pH scale.

Universe   All the stars, galaxies and space itself.

unsaturated   A molecule with at least one double bond between carbon atoms.

urea   A nitrogen-rich substance in urine.

urine   Fluid produced by the kidneys, which contains waste materials from the body, water and salts.

vacuum   A place where there is no matter at all, e.g. space.

variation   Differences between characteristics in different organisms.

(H) vasoconstriction   Narrowing of the blood vessels (capillaries).

(H) vasodilation   Widening of the blood vessels (capillaries).

vector   Organism that transfers a pathogen from one person to another, such as Anopheles mosquito which spreads the protozoan that causes malaria when it bites a human.

vertebra   Small bone in the backbone of a vertebrate (plural: vertebrae).

vertebrate   Animal with a backbone.

virus   A particle that can infect cells and cause the cells to make copies of the virus, such as the influenza virus.

viscosity   How thick or runny a liquid is. Low viscosity is very runny, high viscosity is thick.

visible light   Electromagnetic waves that can be detected by the human eye.

viviparous   Mother gives birth to live young, as in mammals.

voltage   A measure of the amount of energy transferred by a current.

watt (W)   The unit for measuring power. 1 watt = 1 joule of energy transferred every second.

wave power   Generating electricity using the movement of water waves.

wave speed   The distance that a wave travels in one second.

wavelength   The distance between a point on one wave and the same point on the next wave.

white dwarf   A very dense star that is not very bright. A red giant turns into a white dwarf.

wind turbines   A kind of windmill that generates electricity when moving air turns a set of turbine blades.

word equation   Description of a reaction using the names of the reactants and products.

X-rays   Electromagnetic radiation that has a shorter wavelength than UV but longer than gamma rays.

α Abbreviation for alpha particles.

β Abbreviation for beta particles.

γ Abbreviation for gamma rays.

The Periodic Table of the Elements

1	2											3	4	5	6	7	0
							1 H hydrogen 1										4 He helium 2
7 Li lithium 3	9 Be beryllium 4											11 B boron 5	12 C carbon 6	14 N nitrogen 7	16 O oxygen 8	19 F fluorine 9	20 Ne neon 10
23 Na sodium 11	24 Mg magnesium 12											27 Al aluminium 13	28 Si silicon 14	31 P phosphorus 15	32 S sulfur 16	35.5 Cl chlorine 17	40 Ar argon 18
39 K potassium 19	40 Ca calcium 20	45 Sc scandium 21	48 Ti titanium 22	51 V vanadium 23	52 Cr chromium 24	55 Mn manganese 25	56 Fe iron 26	59 Co cobalt 27	59 Ni nickel 28	63.5 Cu copper 29	65 Zn zinc 30	70 Ga gallium 31	73 Ge germanium 32	75 As arsenic 33	79 Se selenium 34	80 Br bromine 35	84 Kr krypton 36
85 Rb rubidium 37	88 Sr strontium 38	89 Y yttrium 39	91 Zr zirconium 40	93 Nb niobium 41	96 Mo molybdenum 42	[98] Tc technetium 43	101 Ru ruthenium 44	103 Rh rhodium 45	106 Pd palladium 46	108 Ag silver 47	112 Cd cadmium 48	115 In indium 49	119 Sn tin 50	122 Sb antimony 51	128 Te tellurium 52	127 I iodine 53	131 Xe xenon 54
133 Cs caesium 55	137 Ba barium 56	139 La* lanthanum 57	178 Hf hafnium 72	181 Ta tantalum 73	184 W tungsten 74	186 Re rhenium 75	190 Os osmium 76	192 Ir iridium 77	195 Pt platinum 78	197 Au gold 79	201 Hg mercury 80	204 Tl thallium 81	207 Pb lead 82	209 Bi bismuth 83	[209] Po polonium 84	[210] At astatine 85	[222] Rn radon 86
[223] Fr francium 87	[226] Ra radium 88	[227] Ac* actinium 89	[261] Rf rutherfordium 104	[262] Db dubnium 105	[266] Sg seaborgium 106	[264] Bh bohrium 107	[277] Hs hassium 108	[268] Mt meitnerium 109	[271] Ds darmstadtium 110	[272] Rg roentgenium 111							

Elements with atomic numbers 112–116 have been reported but not fully authenticated

Key

relative atomic mass
Atomic symbol
name
atomic (proton) number

* The lanthanoids (atomic numbers 58–71) and the actinoids (atomic numbers 90–103) have been omitted.
The relative atomic masses of copper and chlorine have not been rounded to the nearest whole number.

Physics formulae

There are a number of formulae that you will be expected to be able to use in your physics examination for Unit P1. However, you do not need to learn them by heart. You will be a given a formulae sheet in the examinations which will contain all the formulae from the unit. They are also shown below.

Specification statement	Equation
1.15	The relationship between wave speed, frequency and wavelength: wave speed (metre/second, m/s) = frequency (hertz, Hz) × wavelength (metre, m) $$V = f \times \lambda$$ The relationship between wave speed, distance and time: Wave speed (metre/second, m/s)= distance (metre, m)/time (second, s) $$V = \frac{x}{t}$$
5.3	The relationship between electric power, current and potential difference: electrical power (watt, W) = current (ampere, A) × potential difference (volt, V) $$P = I \times V$$
5.17	Calculating the cost of the electricity: cost = power (kilowatts, kW) × time (hour, h) × cost of 1 kilowatt-hour (kW h)
5.21	The relationship between power, energy and time: power (watt, W) = energy used (joule, J) / time taken (second, s) $$P = \frac{E}{t}$$
6.5	The term efficiency calculated from $$\text{efficiency} = \frac{\text{(useful energy transferred by the device)}}{\text{(total energy supplied to the device)}} \times 100\,\%$$

Index

Published by Pearson Education Limited, a company incorporated in England and Wales, having its registered office at Edinburgh Gate, Harlow, Essex, CM20 2JE. Registered company number: 872828

Edexcel is a registered trademark of Edexcel Limited

Text © Pearson Education Limited 2011

The rights of Mark Levesley, Richard Grime, Miles Hudson, Penny Johnson, Sue Kearsey, Damian Riddle, Nigel Saunders, Pauline Anderson, Mark Grinsell, Mary Jones, Ian Roberts and David Swann to be identified as authors of this work have been asserted by them in accordance with the Copyright, Designs and Patents Act 1988.

First published 2011

10 9 8 7 6 5 4

British Library Cataloguing in Publication Data
A catalogue record for this book is available from the British Library

ISBN 978 184690 889 7

Illustrated by ODI
Picture research by Kay Altwegg and Charlotte Lippmann
Printed in the UK by Scotprint

Acknowledgements: The publisher would like to thank the following people for their invaluable help in the development and trialling of this course.

Steven Rowe, Graham Hartland, David French and students at Tomlinscote School; Alex Dawes and students at the Jewish Free School; Suzanne Mycock, Sandra Fox and students at Chelmer Valley High School; David Liebeschuetz, Richard Brock, Elizabeth Andrews and the science team at Davenant Foundation School; Peter Bowen-Walker; Carol Chapman; Gary Gibbons; Gaynor Kendall; Ben Lovick; Esther Ruston; Rupert Turpin and Stephen Winrow-Campbell

We are grateful to the following for permission to reproduce copyright material:

Tables: Graph on page 59 adapted from Trends in the Prevalence and Incidence of Diabetes in the UK - 1996 to 2005, 3 (Elvira Lujan Masso-Gonzalez, Saga Johnansson, Mari-Ann Wallander and Luis Alberto Garcia-Rodriguez 2009), Copyright (c) British Medical Journal (BMJ) 2009; Table on page 71 adapted from http://upload.wikimedia.org/wikipedia/commons/9/94/Cancer_smoking_lung_cancer_correlation_from_NIH.svg, Copyright (c) National Institutes of Health, a biomedical research agency of the U.S. federal government; Graph on page 71 from Statistics on Smoking: England, 2009, Copyright (c) 2010, Re-used with permission of The Health and Social Care Information Centre. All rights reserved; Graph on page 86 from figure I. Estimated and projected population of the world by projection variant, 1950-2050, p7, World Population Prospects: The 2002 Revision, 26 February 2003, ESA/P/WP.180, World Population to 2300 (c) United Nations, 2004. Reproduced with permission; Table on page 86 from http://knol.google.com/k/effects-of-industrial-agriculture-of-crops-on-water-and-soil#, Copyright (c) International Fertilizer Industry Association (IFA); Table on page 95 adapted from Nitrogen fixation in Agriculture, Forestry, Ecology and the Environment, 25-42 31 (M. Hungria, J.C. Franchini, R.J. Campo and P.H. Graham 2005)

The author and publisher would like to thank the following individuals and organisations for permission to reproduce photographs:

(Key: b-bottom; c-centre; l-left; r-right; t-top)

Alamy Images: Adrian Sherratt 180, Andrew Darrington 96tl, Andrew Twort 136t, Angela Hampton Picture Library 184tr, Bill Grant 158, Cathy Melloan 110, D. Hurst 172t, David Forster 154l, David Hoffman Photo Library 184c, David Hosking 83r, David J. Green - financial 224cl, Evan Bowen-Jones 125, Finnbarr Webster 67, GFC Collection 40, Global Warming Images 172c, Graeme Peacock 153, ImageState 156tr, inga spence 66t, INTERFOTO 68b, 148tr, INTERFOTO 68b, 148tr, jaileybug 5b, 272b, Janine Wiedel Photolibrary 154t, Jason Bye 173, Keith Erskine 148l, Leslie Garland Picture Library 166b, 214b, louise murray 246, Lyroky 256, Martin Hughes-Jones 92b, Martin Shields 60b, Mary Evans Picture Library 142, Mayday 156b, mediablitzimages (al) Limited 80l, Melinda Podor 104, Michael Dunlea 261, Nigel Cattlin 90cl, 105, Nigel Cattlin 90cl, 105, Olaf Doering 147tr, P Cox 270, Patrick Lynch 60t, Photo Japan 264t, PhotoAlto 52t, Pictorial Press Ltd 240, Realimage 262t, sciencephotos 160cl, Scott Warren 175, Simon Hadley 263l, sportsphotographer.eu 68t, Susan E. Degginger 126b, The Marsden Archive 242t, Tim Whitby 116-117, Tommy Trenchard 248, Top-Pics TBK 97tr, Tracy Hebden 156b, Trevor Smithers ARPS 96tr, Ulrich Doering 97tl, Victoria Coombs 155tr, William Leaman 96tc, Wolfgang Pölzer 26-27; aviation-images.com: P Jarrett 182; Blackthorn Arable Ltd: 64c; Bridgeman Art Library Ltd: Gassed, an oil study, 1918-19 (oil on canvas), Sargent, John Singer (1856-1925) / Private Collection / Photo © Christie's Images 144t; British Sugar plc: 172b; Construction Photography: David Stewart-Smith 162t; Corbis: Annette Soumillard / Hemis 253tl, Bettmann 157b, Gavin Wickham; Eye Ubiquitous 223l, ILYA NAYMUSHIN / Reuters 152b, KAI PFAFFENBACH / Reuters 140t, Louise Gubb 120t; Courtesy of Philips Consumer Electronics: 266t; Crown Copyright/MOD: Reproduced with the permission of the Controller of Her Majesty's Stationery Office 244b; CTBTO: 243t; DK Images: Dorling Kindersley 30b, Harry Taylor 127br, Kim Taylor 268; Don Dixon: 237b; EnviroMission Ltd: www.tcctaylor.com 252t; ESA: C. Carreau 209, CNES / Arianespace - Optique vidéo du CSG, L. Boyer 174tl; Mary Evans Picture Library: Illustrated London News 214tl; Manchester Evening News: Chris Gleave 90t; EVN: 254t; FLIR: 224bl, 224br; FLPA Images of Nature: Nigel Cattlin 65b; Food Features: 266b, Steve

Moss 65t; Gemini Observatory/Association of Universities for Research in Astronomy: 215; GeoScience Features Picture Library: 120b, 196tl; Getty Images: AFP 38t, China Photos 141l, Denis Doyle 252b, Dick Swanson / Time Life 64t, JIJI PRESS / AFP 32t, Peter Dazeley 56, Roy Stevens / / Time Life Pictures 222t, Still Pictures 168t; Hentis Rail: Antony Henley 260; Image Quest Marine: 38br; iStockphoto: acilo 284, Alena Dvorakova 138t, Alexander Raths 18, Antti-Pekka Lehtinen 5t, 29br, Armin Hinterwirth 31, BanksPhotos 19t, 212b, Caziopeia 264b, Christopher Jones 187b, Clayton Hansen 112t, Dan Eckert 194, Egidijus Skiparis 98, Elementalimaging 176l, Eric Isselee 101r, Heinrich Volschenk 138br, JoLin 198, Laurent davoust 8t, Matauw 108, parameter 19b, 212t, Scott Hirko 101l, Simon Podgorsek 169, stevedangers 8b, Tamara Murray 6tl, 191l, Tomas Bercic 122b, Tyler Boyes 6tc, 191c, Volker Rauch 196c, Volker Rauch 196c, Yuriy Sukhovenko 230r, zoomstudio 122t; Jonny Keeling: 34t; Sue Kearsey: 90bl, 90br; Kobal Collection Ltd: 20TH CENTURY FOX / THE KOBAL COLLECTION 220, UNIVERSAL TV 226tr; Laurence Kemball-Cook, Pavegen Systems: www.pavegen.com: 258; Mark Levesley: 41l, 41r, 62, 88t, 88cl, 94t; LKAB: 152t; Martyn F. Chillmaid: 66b; NASA: Akira Fujii and Infrared Astronomical Satellite 221, ESA / NASA / JPL / University of Arizona 118t, JPL-Caltech 218br, NASA / JPL / Space Science Institute 118bl, NASA / The Hubble Heritage Team / STScI http: / / www.stsci.ed...] / AURA [http: / / www.aura-ast...]) 118br; NASA: 236cl, HEASARC 232bl, NASA / WMAP Science Team 241, NASA Jet Propulsion Laboratory (NASA-JPL) 228t, NASA Langley Research Center (NASA-LaRC) 234c, NASA, ESA, M. Robberto (Space Telescope Science Institute / ESA) and the Hubble Space Telescope Orion Treasury Project Team 228c; Natural History Museum Picture Library: 30t; Nature Picture Library: Adrian Davies 91t, HERMANN BREHM 33l; Norfolk Skyview: Mike Page 216; Oscar & Dehn Ltd: 223r; Pearson Education Ltd: Digital Stock 78, Digital Vision 38l, 187t, Fancy. Veer. Corbis 200t, 202, Gareth Boden 141r, Photodisc 159t, Photodisc. C Squared Studios 36l, Photodisc. John A. Rizzo 130tr, Photodisc. Photolink. E. Pollard 84cl, Photodisc. Photolink. S. Meltzer 155tl, Photodisc. Siede Preis Photography 127t, Photodisc. StockTrek 232t, Trevor Clifford 6tr, 126c, 132l, 134tl, 138bl, 146b, 164b, 191r, 196r, Trevor Clifford 6tr, 126c, 132l, 134tl, 138bl, 146b, 164b, 191tr, 196r; Trevor Hill 184b; Pearson Education Ltd: Trevor Clifford 136b, 185; Photolibrary.com: Imagesource 73r, UpperCut Images 72b; Photoshot Holdings Limited: NHPA / Dave Watts 121; Press Association Images: APTN 250; Reproduced by permission of The Royal Society of Chemistry: Visual Elements – Chemical Data – Radium (http: / / www.rsc.org / chemsoc / visualelements / PAGES / data / radium_data.html) 226b; Reuters: Adeel Halim 147tl, Ho New 50t, 80r, Maldives Government 170, Osman Orsal 36r, Steve Saywell / Royal Navy 146t, Will Burgess 222b; Rex Features: 166t, DAVID HARTLEY 76, Patrick Frilet 124, RICHARD CRAMPTON 178, Solent News 271, Top Photo Group 274; Robert Harding World Imagery: Cosmo Condina 208; Rothamsted Research Ltd.: 58; Science Photo Library Ltd: 70t, 176t, 219l, ANATOMICAL TRAVELOGUE 75l, ANDREW LAMBERT PHOTOGRAPHY 130l, 144b, 179, ANDREW LAMBERT PHOTOGRAPHY 130l, 144b, 179, ARNOLD FISHER 148cl, 148br, BEN JOHNSON 148bl, BIOPHOTO ASSOCIATES 148cr, BJORN RORSLETT 219r, CCI ARCHIVES 218bl, CHRIS BUTLER 236bl, CNRI 73cl, CONEYL JAY 58c, CORDELIA MOLLOY 75r, CUSTOM MEDICAL STOCK PHOTO 82, DAVID NUNUK 160t, DAVID PARKER 247t, DR KEN MACDONALD 85cl, DR MORLEY READ 92t, DR P. MARAZZI 48t, EYE OF SCIENCE 46c, 157t, HENRY GROSKINSKY, PETER ARNOLD INC 269t, J.C. REVY, ISM 4b, J.C. REVY, ISM 4b, 28b, JAMES KING-HOLMES 44t, JIM AMOS 129, JOHN READER 126t, L. WILLATT, EAST ANGLIAN REGIONAL GENETICS SERVICE 42b, MARTIN LAND 148bc, MARTIN M. ROTKER 73tl, MATT MEADOWS, PETER ARNOLD INC. 70cl, MAX ALEXANDER / LORD EGREMONT 210, MEDIMAGE 70cr, MERLIN TUTTLE / BAT CONSERVATION INTERNATIONAL 42t, NASA 234b, PASQUALE SORRENTINO 159t, PEGGY GREB / US DEPARTMENT OF AGRICULTURE 244tr, PETER MENZEL 86t, POWER AND SYRED 42c, RIA NOVOSTI 255, ROYAL OBSERVATORY, EDINBURGH / AATB 292, SAM OGDEN 145, SHEILA TERRY 218t, SHEILA TERRY 218t, SHELIA TERRY 238, SIMON FRASER / RVI, NEWCASTLE-UPON-TYNE 46b, SINCLAIR STAMMERS 77, 150bl, SINCLAIR STAMMERS 77, 150bl, ST MARY'S HOSPITAL MEDICAL SCHOOL 79, STEVE GSCHMEISSNER 84cr, WALLY EBERHART, VISUALS UNLIMITED 94b, Y. SOULABAILLE / EURELIOS 206-207, ZEPHYR 159b; Shutterstock.com: Antonio V. Oquias 230l, Timothey Kosachev 13; SkinzWraps,Inc.: 5br, 272t; SODIS: 276; The Ohio State University: 174tr; The Ohio State University Radio Observatory and the North American AstroPhysical Observatory (NAAPO).: 235; Thinkstock: BananaStock 74t, Comstock 232br, Hemera 46l, 84t, Hemera 48l, 84t, Hemera 46l, 84t, iStockphoto 4t, 28t, 33r, 34cl, 34cr, 34b, 74br, 83l, 85t, 127l, 128, 132, 134tr, 162br, 164t, 168l, 224tr, 242b, 254b, 285, iStockphoto 4t, 28t, 33r, 34cl, 34cr, 34b, 74br, 83l, 85t, 127l, 128, 132, 134tr, 162br, 164t, 168l, 224tr, 242b, 254b, 285, iStockphoto 4t, 28t, 33r, 34cl, 34cr, 34b, 74br, 83l, 85t, 127l, 128, 132, 134tr, 162br, 164t, 168l, 224tr, 242b, 254b, 285, iStockphoto 4t, 28t, 33r, 34cl, 34cr, 34b, 74br, 83l, 85t, 127l, 128, 132, 134tr, 162br, 164t, 168l, 224tr, 242b, 254b, 285, iStockphoto 4t, 28t, 33r, 34cl, 34cr, 34b, 74br, 83l, 85t, 127l, 128, 132, 134tr, 162br, 164t, 168l, 224tr, 242b, 254b, 285; Thomas G. Ranney, Ph.D.: 64b; TopFoto: 243b; University of North Carolina: Professor Joseph DeSimone 46t; Wicab,Inc. www.wicab.com: 54; www.tidalstream.co.uk: 253tr; www.unece.org/trans/danger/publi/ghs/pictograms.html: Reproduced with the kind permission of the secretariat of the United Nations Economic Commission for Europe 140cl, 140cr, 140bl, 140br

All other images © Pearson Education

Every effort has been made to contact copyright holders of material reproduced in this book. Any omissions will be rectified in subsequent printings if notice is given to the publishers.

Disclaimer: This material has been published on behalf of Edexcel and offers high-quality support for the delivery of Edexcel qualifications. This does not mean that the material is essential to achieve any Edexcel qualification, nor does it mean that it is the only suitable material available to support any Edexcel qualification. Edexcel material will not be used verbatim in setting any Edexcel examination or assessment. Any resource lists produced by Edexcel shall include this and other appropriate resources.

Copies of official specifications for all Edexcel qualifications may be found on the Edexcel website: www.edexcel.com